T0171777

GeoGuide

Series Editors

Wolfgang Eder, University of Munich, Munich, Germany

Peter T. Bobrowsky, Geological Survey of Canada, Natural Resources Canada, Sidney, BC, Canada

Jesús Martínez-Frías, CSIC-Universidad Complutense de Madrid, Instituto de Geociencias, Madrid, Spain

Axel Vollbrecht, Geowissenschaftlichen Zentrum der Universität Göttingen, Göttingen, Germany

The GeoGuide series publishes travel guide type short monographs focussed on areas and regions of geo-morphological and geological importance including Geoparks, National Parks, World Heritage areas and Geosites. Volumes in this series are produced with the focus on public outreach and provide an introduction to the geological and environmental context of the region followed by in depth and colourful descriptions of each Geosite and its significance. Each volume is supplemented with ecological, cultural and logistical tips and information to allow these beautiful and fascinating regions of the world to be fully enjoyed.

More information about this series at http://www.springer.com/series/11638

Mike Searle

Geology of the Oman Mountains, Eastern Arabia

 Springer

Mike Searle
Department of Earth Sciences
Oxford University
Oxford, Oxfordshire, UK

ISSN 2364-6497   ISSN 2364-6500   (electronic)
GeoGuide
ISBN 978-3-030-18452-0   ISBN 978-3-030-18453-7   (eBook)
https://doi.org/10.1007/978-3-030-18453-7

This Springer imprint is published by the registered company Springer Nature Switzerland AG
The registered company address is: Gewerbestrasse 11, 6330 Cham, Switzerland

This book is dedicated to my late parents, Geoff and Pauline Searle, with grateful thanks for the wonderful life we had in Oman in the early days. It is also dedicated to all those like-minded conservationists, naturalists, and geologists, both local Omanis and Emiratis as well as foreigners, who are working towards the preservation of the rich natural heritage of this amazing and wonderful country.

Foreword

Oman and the UAE possess several geological wonders that capture the evolution history of the Arabian Plate and contribute significantly to our understanding of key Earth dynamic processes. Between the Oman Mountains in the north and the Dhofar Mountains in the south, many geological windows and terrains expose a wide range of rock formations and provide astonishing geological landscapes and phenomena that are only present in few places across the globe. These include the Proterozoic glaciation cycles, the Palaeozoic tectonic events and climate history in Gondwana, the Permian–Mesozoic shelf carbonates that contain large oil and gas reserves in the interior, and the classic Semail Ophiolite thrust sheet of Tethyan Ocean crust and mantle, the world's largest and best-preserved slice of oceanic lithosphere emplaced onto the continental margin.

Professor Mike Searle is certainly one of the best people to write about the geological record of the Oman Mountains, having actively done research in this area since 1978. Mike's father worked in Petroleum Development Oman from 1967 to 1979, and his mother was the Reuters correspondent. Hence, he has had a strong personal attachment to the country, its natural history, geology and people, and he has visited both Oman and UAE almost every year since 1968 first as a schoolboy and student, later working in the mountains for his PhD, and more recently studying research projects in his position as professor of Earth Sciences at Oxford University.

The Geological Society of Oman has been leading the way in Oman to get many of the critically endangered and best geological sites preserved as GeoParks and Sites of Special Scientific Interest (SSSI) established into law. The governmental authorities in Oman are currently enforcing laws that ensure the preservation. Mike has been working closely with the Geological Society of Oman to emphasise the message. He has suggested more than 50 unique geological sites that should be preserved. As he stresses in this book, development is necessary but must be done in suitable places, not destroying the natural environment. It is crucial that developers and planners interact with geologists and naturalists before any more critically important sites are lost under concrete. Mike has always raised his concerns about the preservation of the geological sites in Oman and the UAE, as both countries have seen enormous and rapid development during the last 50 years. Urban development is spreading with astonishing speed. Artificial islands have been built in the UAE from rocks quarried in the mountains. Population growth in both countries is increasing quite rapidly, and the countries are looking at ways to diversify the economies. It is now more important than ever to preserve the natural environment.

Throughout this revealing book, Mike will share with you some of the most important geological wonders in Oman and the UAE, describing the main geology sites from the Musandam Mountains in the north to the Eastern Mountains along the Indian Ocean coast. The book gives a brief introduction about the main rock sequences in the region and their environment of deposition and tectonic setting. Being closely attached to the Ophiolite rocks and the related obducted and sub-ducted units, Mike delegates a separate chapter to describe these sequences and their uniqueness in the world. The book also covers the foreland desert region including the Liwa sands, the Wahiba Sand Sea in the Sharqiyah region and the Rub al-Khali desert and great sand dune country of the interior. His call for conserving the geological wonders seals the last chapter of the book; as indeed, this is one of the main reasons Mike has written this great book.

Muscat, Oman   Dr. Mohammed al-Kindy
                                                          Earth Sciences Consultancy Centre

Preface

Arabia is one of the most enigmatic and unknown continents on Earth. The vast Rub al-Khali desert, known as the Empty Quarter, occupying most of the landmass of Arabia was one of the last of the great continental extreme environments to be explored. Richard Burton, William Palgrave, Charles Doughty, Gertrude Bell and Freya Stark were a few of the redoubtable explorers who ventured into the vast desert wasteland of the Rub al-Khali, in what is now Saudi Arabia and the frankincense coast of Dhofar, Hadramaut and the Magreb (Yemen) in south Arabia. Although local bedouin certainly made long-distance camel treks across parts of Arabia, the first recorded north–south crossing of the Empty Quarter was achieved by Bertram Thomas in 1930–1931, described in his book '*Arabia Felix*'. This amazing journey was followed closely by Harry St-John Philby in 1932, who travelled west to east across the great desert. Wilfred Thesiger's name is imprinted on the exploration of Arabia with his 20 years of travels, almost entirely on foot and with camels, exploring vast tracts of unexplored country mainly in eastern Arabia and his two great crossings of the continent in 1946–1949. His book '*Arabian Sands*' is one of the classic travel books of all time. Thesiger was instrumental in discovering Fahud the huge folded mountain range where oil was first discovered in Oman in the early 1960s. Now over half the world's oil and gas is produced from wells along the Arabian Gulf and eastern Arabia, from Iran, Iraq, Kuwait and Saudi Arabia through Qatar and the United Arab Emirates to Oman.

In eastern Arabia, the mountains of Oman are a spectacular and impressive range, unlike any other. These desert mountains contain three large limestone massifs, the Musandam peninsula, Jebel Akhdar and Saih Hatat, that expose the stratigraphic sections of the oil-bearing strata drilled in the deserts to the west and south. The mountains rise to just over 3000 metres and have numerous deeply eroded canyons with perennial flowing streams, tiny date palm villages with terraced fields and intricate *falaj* irrigation systems, ancient juniper woodlands and spectacular walls of rock. The coastline of Oman and the eastern United Arab

Emirates (UAE) is stunning, with white-sand beaches, rocky coves and offshore coral reefs. The entire Musandam peninsula has tilted to the east and northeast during the Quaternary following the onset of continental collision, and the drowned coastline has spectacular deep fjords surrounded by towering limestone cliffs.

The Oman mountains are, however, geologically famous for having the world's largest ophiolite complex, a thrust slice of oceanic crust and upper mantle formed 96–95 million years ago, and emplaced onto the Arabian continental margin during the Late Cretaceous between 95 and 70 million years ago. Oman is a geological paradise, not only for petroleum geologists but also for oceanographers, petrologists, structural geologists, palaeontologists and marine scientists. The Oman ophiolite played a major role in the geological studies that resulted in the theory of plate tectonics, proposed in the 1960s. The Shell team of geologists who first mapped the entire mountain range during the winters of 1966–1968, led by Ken Glennie, produced a spectacular and very accurate geological map, and they first defined in detail the stratigraphy and complex structure of the Oman mountains. The Shell Memoir published in 1974 laid the foundations for all future work in Oman. Tectonic processes of how plates collide, how narrow slivers of continental crust could have been subducted to great depths ($\sim$ 100 km) and then exhumed and how oceanic plates are emplaced onto continental margins have been mainly solved by mapping and studying the rocks of the Oman–UAE mountains.

The Trucial coast, also known as the pirate coast, centred around the coastal trading ports of Abu Dhabi, Dubai, Sharjah, Ajman, Umm al-Quwain and Ras al-Khaimah, was a sea-based economy reliant mainly on local pearl diving, and dhow-based trade with Basra in Iraq, the Persian coast and India. Following long periods of coastal piracy along the Gulf, the Trucial coast became a British Protectorate in 1819, a situation that lasted until 1971. Six sheikdoms, Abu Dhabi, Dubai, Sharjah, Ajman, Umm al-Quwain and Fujairah, united to become the federation of the United Arab Emirates (UAE) with Sheik Zayed bin Sultan Al Nahyan as its head. The largely mountainous sheikdom of Ras al-Khaimah, which includes the western part of the Oman-controlled Musandam peninsula, joined a year later. The drawing of the borders around each sheikdom resulted in a complex map where each emirate has three or four separate parts of territory. A small enclave of Oman (Madhah) within Fujairah even has an even smaller enclave of Fujairah (one village of about 20 houses) within it. UAE has extremely large oil reserves, mostly within the Sheikdom of Abu Dhabi. The development of both Oman and the United Arab Emirates, especially Dubai, Abu Dhabi and Sharjah, is one of the fastest and most extreme social, political and material changes ever seen. These cities and countries have risen from dusty, desert wastelands and poor fishing villages to skyscraper cities in only 20 years.

The history of oil exploration in Arabia goes back more than 60 years and is a fascinating story of tribal uprisings, political manoeuvrings by the oil companies and western powers, together with some crucial geological judgments. The 1949–1955 Buraimi Dispute between the house of Saud in Saudi Arabia, Sultan Said bin Taimur of Oman, and the sheiks of the Trucial States, notably Sheik Zayid of Abu Dhabi, is an intriguing tale that resulted in the political borders of eastern Arabia finally being drawn in the 1950s. Following early oil prospecting around the eastern deserts of Arabia, some particular mountain structures around Buraimi and Al Ain, notably the spectacular large anticline structure of Jebel Hafit, were of special interest to geologists. In 1952, Saudi Arabia sends a small force to occupy part of the Oman-controlled Buraimi oasis, but in 1955 the British sent a detachment of the Trucial Oman Scouts to expel them and retain Buraimi for Oman. The 'Jebel war' in Oman (1954–1959) between the forces of Sultan Said bin Taimur of Oman and the self-styled 'Lord of the Green Mountain' Sulaiman bin Himyar Al Ryami and the Imam, Ghalib bin Ali al Hinai with Saudi backing, was won by the Sultan's forces with British SAS backing, and resulted in the settling of all northern Oman into the present country of Oman. The Dhofar rebellion in the south (1963–1976), fought between the Sultan's forces and the Aden-based communists supported by the old Soviet Union in Yemen, resulted in a decisive victory for the Sultan's forces and effectively uniting all coastal areas of eastern Arabia and the tribal areas to the south and east of the Rub al-Khali into the Sultanate of Muscat and Oman.

Up to 1967, Oman was one of the most backward countries in the world with no roads, schools or hospitals in the entire country, despite having a sea-faring tradition that extended to Omani Arab colonies in Baluchistan and Makran, along the Malabar coast of India and along the east coast of Africa. Omani seafarers and traders established several trading posts, notably on the islands of Zanzibar and Pemba in what is now Tanzania, as far back as 1689. Up to 1964, the country was known as the Sultanate of Muscat, Oman and Zanzibar, but in April 1964 Zanzibar was united with Tanganyika to become Tanzania. Muscat and Zanzibar were centres of the frankincense and spice trade with extensive dhow routes from Muscat along the coasts of the Indian Ocean from the Red Sea south to Zanzibar and east to Java and the South China Sea. Oman remained a remote, undeveloped and largely unexplored country up until July 1967 when a palace coup in Salalah by Sultan Qaboos bin Said ousted his father, Sultan Said bin Taimur. From this time on, Oman was set on an incredibly rapid road to change and rapid improvement in the standard of living for all Omanis.

The modern geological exploration of the Oman–UAE Mountains began in earnest in the late 1960s and 1970s. The geological heritage is one of the richest anywhere on earth and great efforts are now being made to preserve many unique geological sites before development takes over. This book describes some, but by no means all, of these geological sites of importance. It is hoped that many of these sites will soon become UNESCO World Heritage sites, GeoParks, or Sites of Special Scientific Importance (SSSI). It is hoped that this book will raise the profile of these world-class sites and preserve them from the bulldozers and diggers that are so widespread across the Middle East today. The preservation of these GeoParks is intimately tied in to nature conservation areas, some already established to help conservation of rare Arabian leopards, Arabian Tahr, Oryz, Turtles and other endangered species. The establishment of GeoParks and nature reserves will also inevitably help preserve the old traditions of Omani, Emirati and Bedouin cultures as well as the wonderful hospitality of the Omani and Emirati people. The ancient ways of the bedouin have been superseded by modern twenty-first century states, but as Wilfred Thesiger so eloquently put it: '*No man can live this life and emerge unchanged. He will carry, however faint, the imprint of the desert, the brand which marks the nomad; and he will have within him the yearning to return, weak or insistent according to his nature. For this cruel land can cast a spell which no temperate clime can match*'.

Oman mountains showing the dark hills of ophiolite around Semail village and large limestone mountains of Jebel Nakhl beyond. Photograph: Mike Searle

Oxford, UK   Mike Searle

Acknowledgements

I am totally indebted to my wonderful parents Pauline and Geoff Searle, now both sadly passed away, who not only brought our family to live in Oman from 1967 to 1979 but instilled a great sense of exploration and wonder in the natural environment in me. We explored the magnificent mountains, coastline and deserts of Oman in days when there were no roads, and the countryside was pristine. I thank our old family friends Diyab al-Amri, George and Diana Wadia, late Don and Elouise Bosch, David Shepherd, and many others who shared our exploration trips in the early days. I continued working in Oman for my Ph.D. thesis from 1977 to 1980, and since then have returned to Oman and UAE almost every year with students and colleagues to work on the remarkable geology. David, Judy, Mateo, Susie and Miguel Willis were the most fantastic hosts and lifelong friends; thanks to them for using their house in Bausher as a base camp for many years.

I made several wonderful trips to the interior and Dhofar with David, Judy, Susie and Mateo Willis, Hadi al-Hikmani, Khalid al-Hikmani, late Ralph Daly, Andrew Spalton, Patrick van Daele, and Roddy Jones, several birdwatching trips with the late Michael Gallagher, and Hanne and Jens Eriksen. Mohammed al-Kindy and Husam al-Rawahi of PDO, and Robert Whitcombe and Claire Bond of Atkins, Oman, were always extremely helpful, especially in our efforts to get the GeoParks into Oman law. Nigel and Shane Winser at the Royal Geographical Society in London have always been immense supporters of all things Omani. Thanks to Said bin Jabber for organising a wonderful camel trek across the Wahiba Sands in 1985. In UAE, I thank Mohammed Ali of the Petroleum Institute, now Khalifa University, for great hospitality in Abu Dhabi over many years, and also to Aisha Al-Hajri, and Ahmed the goat farmer in Masafi.

From Petroleum Development, Oman (PDO), I would like to thank Ken Glennie, Pete Jeans, George Band, Alan Heward, Bruce Levell, Craig Harvey, Marcus Hollanders, Mohammed al-Kindy, Husam al-Rawahi, Henk Droste, Mia van Steenwinkle, Ali Al-Lazki, Abdulrahman Al-Harthy and Patrick van Daele for accompanying us on trips, organising permits, and general enthusiasm and interest for all things geological. Thanks also to the Managing Director of PDO, Raoul Restucci, for supporting our GeoParks initiative and promoting our plans to the government.

From the academic side, I would like to extend my great thanks to Hilal al-Azri, late Ishmael El Boushi, late Ian Gass, Bob Coleman, Cliff Hopson, John Smewing, Adrian Lewis, Steve Lippard, Alastair Robertson, David Cooper, Tony Watts, Mohammed Ali, Hugh Rollinson, Ian Alsop, Philippe Agard, Randall Parrish, Brad Hacker, Matt Rioux, Josh Garber, Nasser Al Rizeiqi and Andrew Goudie. Special thanks to Moujahed Al-Husseini, chief editor of GeoArabia, Nestor Buhay and Joerg Mattner (GeoTech, Bahrain) for their great support of our work, and for permission to reproduce several figures from the journal GeoArabia.

My Ph.D. (D.Phil.) and Master's students at Oxford University have done a lot of hard work, and I extend my thanks to all of them: Halcyon Martin, Jon Cox, Clare Warren, Sanat Aidebayev, Tom Jordan, Robbie Cowan, Alan Cherry, Charles Cooper, Sam Cornish, Tyler Ambrose, Brook Keats, Simone Pilia, Lauren Kedar, Adrian White, Ben Tindal and Tom Reeshemius. I also extend my thanks to Dr. Salman Mohammed Al-Shidi of the Ministry of Oil and Gas, Mohammed Saleh Al Fanah Al Arimi, Director of Geological Research and Heritage Department, Dr. Ali Salim Ali Al-Rajhi, Director General of Geological Survey, and Malik Al-Hinai, Director of Bait al Baranda for their help and support.

From the climbing side, I would like to thank Martin Stephens, Neil Graham, Peter King, Dave Mithen, Dana Coffield and Nick Groves for our wonderful climbs on Jebel Misht and around the Jebel Akhdar massif. Tony and Di Howard organised three years of exploration, preparing the first version of our climbing, trekking, caving and exploration guides to the Oman mountains. Thanks to them and also to Alec MacDonald, Bernard Domenech, Mario and Freni Verin.

Many thanks to Bruce Levell, Alan Heward, Mohammed al-Kindi, David Cooper and Mohammed Ali for comments on individual chapters, and to Mohammed al-Kindi for writing the preface. I would like to thank Judy and David Willis for years of hospitality and friendship in Bausher, and Henk Droste and Mia van Steenwinkel for hospitality in Qurum, as well as their encyclopaedic knowledge of Oman. I thank my editor Annett Buettner and Sanjievkumar Mathiyazhagan of Springer publishers for their expert help with editing and production.

Contents

About the Author

Mike Searle is Professor of Earth Sciences at the University of Oxford and a Senior Research Fellow of Worcester College, Oxford. For the last 40 years, he has worked on the geology of the Oman–UAE mountains as well as the geology of the Himalaya, Karakoram and Tibet. He has authored more than 220 peer-reviewed papers, co-edited 5 books of the Geological Society of London, and authored two books, '*Geology and Tectonics of the Karakoram Mountains*' (1991, John Wiley & sons) and '*Colliding Continents: A Geological Exploration of the Himalaya, Karakoram, and Tibet*' (2013; Oxford University Press). He has published many geological maps, notably of the Central Karakoram region, Hunza and Baltoro regions (1991), North Pakistan (1996) and the Mount Everest—Makalu region, Nepal and South Tibet (2003). He worked on the Oman–UAE mountains for his Ph.D. (1980) and has subsequently carried out consulting work for several oil companies, and the Petroleum Institute, Abu Dhabi. He has had ongoing academic research projects throughout Oman and UAE and has visited the country every year from 1980 to the present time.

Acronyms

Industrial and Political

ADNOC	Abu Dhabi National Oil Company
Aramco	Arabian-American Oil Company
BGS	British Geological Survey
BOC	Burma Oil Company
BP	British Petroleum
BRGM	Bureau de Recherches Géologiques et Minieres
CFP	Companié Francaise des Petroles
IPC	Iraq Petroleum Company
MOG	Ministry of Oil and Gas
PDO	Petroleum Development (Oman) Ltd.
PDRY	People's Democratic Republic of Yemen
PI	Petroleum Institute, Abu Dhabi
SAS	Special Air Service (British Army)
SOAF	Sultan of Oman's Armed Forces
SOC	Standard Oil of California
UAE	United Arab Emirates
UN	United Nations
UNESCO	United Nations Educational, Scientific and Cultural Organisation

Geological

EPR	East Pacific Rise
Ga	Billion Years
IAT	Island Arc Tholeiite (Basalt)
IODP	International Ocean Drilling Project

Ka	Thousand Years
Ma (Myr)	Million Years
MAR	Mid-Atlantic Ridge
MIR	Mid-Indian Ocean Ridge
MORB	Mid-Ocean Ridge Basalt
OIB	Ocean Island Basalt
REE	Rare Earth Elements
SSZ	Supra-Subduction Zone Lavas
WPB	Within-Plate Basalt

Part I
Geography, History and Exploration

Arabia: Geography, History and Exploration

1

Arabia is a land of spectacular extremes and natural beauty. These landscapes include the largest sand dunes on Earth in the Rub al-Khali, the lush valleys and marshes of the Tigris and Euphrates in Iraq and Syria, the red desert of Jordan, the great sandstone cliffs of Wadi Rum, the forested hills of the Yemeni highlands and Asir in Saudi Arabia, magnificent coral reefs along the Red Sea and Gulf of Aqaba, the *khareef* monsoon-lashed green sea-cliffs of the frankincense coast in Yemen and Dhofar, the incredible Wahiba Sandsea, the islands and *khors* (*khawrs*) of the Indian ocean coast and Arabian Gulf. However, foremost amongst these natural wonders of Arabia, the unique Northern Oman–United Arab Emirates (UAE) Mountains with their deeply incised *wadis*, perched villages and perennial flowing streams, the stunning beaches and coastline of Oman, the amazing fjords and sea cliffs of the Musandam peninsula, has to be at the top.

1.1 Regional Tectonic Setting

The geography and geomorphology of Arabia is closely linked to its bedrock geology (Fig. 1.1). Geographically, the Arabian continent is bounded by the Mediterranean Sea to the west, the Red Sea to the southwest, the Gulf of Aden and the Indian Ocean to the south and southeast, the Gulf of Oman and Arabian Gulf to the northeast. Geologically the Arabian plate was part of the African plate with a common Precambrian Pan-African basement crust. Arabia as we know it today was formed when Africa and Arabia split apart and oceanic crust was formed along the Red Sea. Active continental rifting and strike-slip faulting occurred along the Levant (Syria, Lebanon, Palestine) margin with initiation of the Dead Sea rift and strike-slip fault zone. New ocean crust is being formed along the axis of the Red

© Springer Nature Switzerland AG 2019
M. Searle, *Geology of the Oman Mountains, Eastern Arabia*,
GeoGuide, https://doi.org/10.1007/978-3-030-18453-7_1

Sea today. Active continental rifting extends south into the Danakil depression in the Afar region, and south of Ethiopia along the East African rift valleys in Kenya, Tanzania and Mozambique. To the north, the Gulf of Suez is a failed rift, and active strike-slip faulting instead extends north along the Gulf of Aqaba and Dead Sea fault though Jordan, Palestine, Lebanon and into Syria. The western and southern margins of Arabia are extensional rifted margins, whereas the northern and northeastern margin of the Arabian plate from southeast Turkey along the Zagros mountains of southwest Iran and east to Baluchistan are compressional, formed in the last twenty million years by the collision of the Arabian plate with central Iran along the Zagros suture zone.

The Arabian plate, moving northeastwards as a result of rifting along the Red Sea, collided with central Iran, closing the intervening Tethyan Ocean sometime around twenty million years ago. Rocks formed in the Tethyan Ocean are preserved as highly deformed remnant ophiolites (fragments of ocean crust and upper

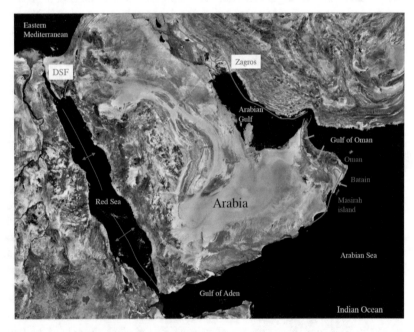

Fig. 1.1 Landsat satellite photo of Arabia

mantle), deep-sea sediments and mélanges along the Zagros suture zone. Oceanic crust and mantle formed in the Tethyan ocean are now preserved as ophiolite sequences at Neyriz and Kermanshah in Iran. The Zagros mountains to the southwest of the suture zone are composed of Arabian plate rocks deformed into a series of spectacular periclinal folds on a huge scale (Fig. 1.2). This is the Zagros fold belt where some 12 km thickness of Phanerozoic sediments has been buckled and crumpled as a result of the plate collision. These sediments overlie a layer of Precambrian Neoproterozoic–Cambrian salt (the Hormuz salt) that has provided the basal detachment to the upper crust folds and thrust faults of the Zagros fold-thrust belt. The salt has been mobilised by the compressional stresses and in places intruded up through the 10–12 km overlying sedimentary package to form spectacular salt domes. These salt domes occur along the entire Zagros mountains and Arabian Gulf region offshore Abu Dhabi in the United Arab Emirates and Oman forelands.

Fig. 1.2 Satellite photo of the Arabian Gulf, Zagros Mountains of SW Iran, Straits of Hormuz, Oman–UAE Mountains and Gulf of Oman

The Arabian Gulf is a flexural foreland basin southwest of the Zagros fold belt. As the Arabian plate collided with Iran, compressional forces caused large-scale folding of the upper crust and underthrusting of Arabian plate lower crust beneath the central Iran plateau. The loading of the Zagros fold and thrust belt caused the Arabian plate to flex down, forming the Arabian (Persian) Gulf. This young foreland basin extends northwest into the Mesopotamian basin along the Euphrates and Tigris rivers in Iraq. In the southeastern Zagros, the continental collision fold and thrust belt is terminated by large strike-slip faults like the Zendan fault, east of the Straits of Hormuz. The suture zone belt of Tethyan oceanic rocks ends and has been faulted out reappearing in the Gulf of Oman and northern Oman Mountains. The southern Makran continental margin in Baluchistan is a large-scale accretionary prism above an active north-dipping subduction zone extending along the Makran trench east towards the Owen Fracture zone in the Arabian Sea.

The transition from a young continent—continent collision zone along the Zagros Mountains of Iran to the continent—ocean collision zone along the Oman–UAE mountains can be seen in the mountainous Musandam peninsula and Straits of Hormuz region where the Arabian Gulf connects with the Gulf of Oman. In the mountains of Musandam the first effects of the continental collision can be seen, with large-scale thrusting and folding during the Late Miocene. To the south along the Oman–UAE mountains the continents have not yet collided and the Gulf of Oman is instead mainly remnant oceanic crust of the Tethyan ocean. The Oman mountains show very different geology (Fig. 1.3). Here, a huge thrust sheet of oceanic crust and upper mantle, the Semail (or Oman) ophiolite has been emplaced southwestwards on top of the Arabian continental margin. These dark coloured ophiolitic rocks give the Oman mountains its unique geomorphology, a rugged, barren sea of spiky black mountains.

Western Arabia is composed of the oldest exposed rocks, the Precambrian (Proterozoic) basement gneisses belonging to the Pan-African orogenic event. Along the Red Sea young volcanic rocks have been erupted during rifting of the Red Sea and overlie these old rocks. The port city of Aden is located in and around the crater of one of these volcanoes. Eastward along the coasts of Yemen and Dhofar, the southern province of Oman, the old Precambrian basement rocks are overlain by Cenozoic sedimentary rocks. These sedimentary rocks show that Arabia was covered by shallow marine conditions up to about 20 million years ago when the peninsula was uplifted and desertification began. The Rub al-Khali dunes of the Empty Quarter now overlay these sedimentary rocks across the middle of Arabia (Fig. 1.4).

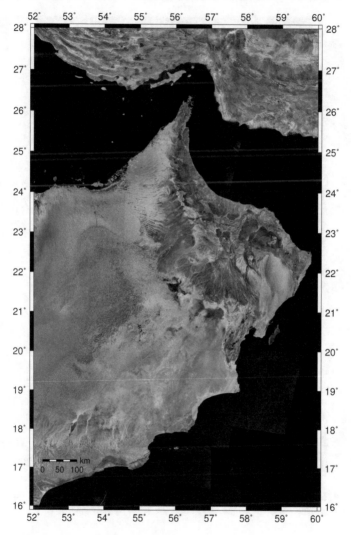

Fig. 1.3 Satellite photo of Eastern Arabia

In southern Oman the basement rocks are exposed again in the Hallaniyat (Kuria Muria) islands and the Marbat plains east of Salalah, overlain by the Paleogene limestones of Jebel Samhan which extend inland to form the desert hinterland. Along the southeast coast of Oman remnant ophiolites, or tectonic

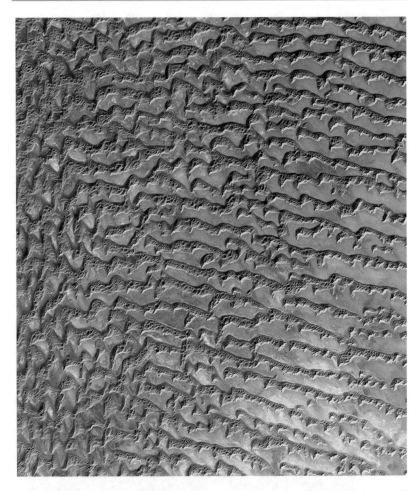

Fig. 1.4 Satellite photo of Rub al-Khali Empty Quarter showing sand dunes with intervening white sabkhas. Image taken from the Terra satellite (EOS)

slices of old Indian Ocean crust and upper mantle are exposed on a few headlands (e.g. Ras Madrakah) jutting out into the Arabian Sea, and on Masirah Island off the coast of central Oman. Oman has two separate mountain ranges, the southern ranges of Jebel Qamar and Jebel Qara behind Salalah extending east to the Jebel Samhan escarpment, and the northern Oman Mountains. The Northern Oman mountains run for over 700 km from the Musandam peninsula in the north to Ras al Hadd, and the northeastern-most tip of Arabia.

1.2 Geomorphology and Vegetation

The central part of Arabia is the great Rub al-Khali, the Empty Quarter, the largest contiguous sand desert anywhere on Earth covering approximately 650,000 km². The fringes of the Rub al-Khali are gravel deserts but the central part consists of massive sandseas with individual dunes reaching over 250 m in height. Rainfall is extremely low, but occasional storms and flash floods have resulted in meandering wadis and there is an extensive underground aquifer system. Temperatures during summer can regularly reach above 50 °C. Interspersed among the sand dunes are patches of white *sabkha*, gypsum and salt flats that become treacherous in the rain. The largest of these is the Umm al Sammim, the Arabic for 'Mother of Poisons', a huge area of quicksand encrusted with salt that collects drainage from many of the large wadis flowing south from the Northern Oman Mountains.

A separate desert sandsea, the Wahiba sands, also known as the Sharkiyah sands, extends for about 180 km north-south and about 80 km east-west covering an area of approximately 12,500 km². It is a unique desert being almost entirely

Fig. 1.5 Wahiba (Sharkiyah) sandsea, central south Oman

Fig. 1.6 Giant seif dunes in the Wahiba sands

composed of large linear seif dunes formed parallel to the onshore northwesterly winds associated with the summer monsoon (Figs. 1.5 and 1.6). Individual dunes can reach over 120 m high. Along the south, older aeolianites are semi-cemented or even fully cemented carbonate fossil dunes that are beautifully exposed along the Indian Ocean coast. The western limit of the Wahiba sands is the large Wadi Andam that drains much of the southern slopes of the eastern part of the North Oman mountains.

Two main mountain ranges occur in eastern Arabia, the Oman–UAE mountains in the northeast and the Jebel Samhan–Jebel Qara–Jebel Qamar mountains in Dhofar along the Indian Ocean coast. These mountains extend westward into the Yemeni districts of Mahra and Hadramaut. In the north of Oman the three major shelf carbonate culminations of Musandam, Jebel al-Akhdar and Saih Hatat make up the highest mountains with the deepest incised wadis. The ophiolite mountains have a unique geomorphology with mountain range upon mountain range of dark, spiky peaks made up of serpentinised peridotites and crustal gabbros and basalts. In the northeast a large limestone plateau region composed of Paleogene limestones makes up most of the mountains in the Qalhat–Sur region. Many caves have been formed including one of the Worlds' largest caves, the Majlis al-Jinn. North of the Oman mountains a desert plain makes up the Batinah coast stretching nearly 200 km from Shinas and the UAE in the northwest to the Qurm region near Muscat in the east. This plain is made up of alluvial sands and gravels eroded off the mountains and is comparatively well vegetated. The deep aquifers are being reduced at an alarming level due to over-development and agricultural needs, such that the once thriving date palms are dying off as salty water inundates the aquifers from the Gulf of Oman.

1.3 Climate

Although most of Arabia has an arid climate, extremes of temperatures are known from summer highs of up to and exceeding 50 °C and winter lows down to freezing. The Oman Mountains along the northeast corner of Arabia reaches 3009 m elevation at Jebel Shams (Jebel al-Akhdar massif; Fig. 1.7). Here snow can very occasionally cover the summit ridge, and the altitude on the higher mountain peaks and at the Saiq plateau provides an ideal temperate climate during the baking

Fig. 1.7 Dawn over the Oman Mountains, view east along Jebel Akhdar, from summit of Jebel Shams (3009 m)

Fig. 1.8 Spring cherry and apricot blossom, Shariyah village, Saiq plateau

summer months. Around the Saiq plateau, cherry and apricot trees grow together with pomegranates, grapes and roses (Fig. 1.8). Most of Arabia receives between 50 and 200 mm rainfall a year, almost all of it falling during the winter months. Sudden tropical rainstorms can occur and when they do, spectacular flash floods cascade down the wadis (Fig. 1.9). After rains the Saiq plateau frequently shows impressive waterfalls falling over steep cliff faces. Travellers in middle Arabia have

Fig. 1.9 Flash floods in Jebel al-Akhdar, Wadi Muaydeen

Fig. 1.10 Flash flood in Wadi Aday

been known to have been stranded on small islands surrounded by a sea of water and mud in the middle of the desert for days following storms like these. In the mountains flash floods are a major hazard with many of the narrow canyons filling up to 100 m depth in a few minutes (Fig. 1.10).

The climate of Arabia is affected mainly by the mid-latitude westerlies during winter months. In summer, strong winds from Asia blowing from northwest to southeast (the *shemal*) can result in duststorms. During summer months (Late June to September) South Arabia is hit by the monsoon-driven easterly winds called the *khareef*, that blow along the Equator. Large ocean upwelling currents off the coast of south Arabia provide ideal conditions for plankton blooms that attract enormous quantities of fish. The monsoon easterlies also cause dense coastal fogs along the southern coast of Yemen and Dhofar with frequent precipitation. The foggy deserts

Fig. 1.11   Satellite photo showing Cyclone *Gonu* over eastern Arabia, 5th June 2007

inland provide the moisture required to sustain a rich wildlife including herds of gazelle and oryx. Past climate changes over thousands, or tens of thousands, years have moulded the landscapes and led to migration of tribes. There is a rich archaeological history from the interior of Dhofar in particular and the ancient lakes fringing the Rub al-Khali.

The western Indian Ocean also spawns intense tropical cyclones which frequently originate in the equatorial latitude and sweep northward brushing the coast of Oman and heading towards the Makran and India. These cyclones are part of the summer monsoon winds that sweep north across India providing intense rainfall and stormy weather to the Indian sub-continent. On 4th June 2007 the strongest tropical cyclone ever recorded in the Indian Ocean, Cyclone *Gonu* developed in the Indian Ocean and moved northwestward hitting the eastern coast of Dhofar in south Oman on 6th June when it weakened to a tropical storm as it travelled inland towards Muscat (Fig. 1.11). Wind speeds up to 270 km/h (165 miles/h) were recorded east of Masirah Island. At Ras al-Hadd a 5.1 m storm surge hit the coast. More than 600 mm of rain fell in 48 hours resulting in a large amount of coastal damage, flash floods, at least 100 deaths and more than $4.2 billion worth of

Fig. 1.12 Flooding effects of Cyclone *Gonu* around Wadi Aday, Qurum, Medinat Qaboos and Muscat

damage to property (Fig. 1.12). In Muscat winds reached more than 100 km/hour, wadis filled up to more than 50 m in places, and huge waves crashed around the coast causing storm surges along the Batinah coast. Even in the desert, major rains caused severe flooding, and several oil-rigs had to stop operations. Two further cyclones, *Phet* in 2010 and *Chapala* in 2015 also resulted in storm surges, intense rain and infrastructure damage.

1.4   History of Exploration

The harshness of the desert landscape of Arabia has long been a deterrent for trade across Arabia so that the earliest records are mainly from seafaring sources. Incursions along the shores of Arabia by Alexander the Great in 326 BC and the Roman Emperor Trajan in 116 AD were the first recorded contacts with western nations, but the Arabs of the Gulf had been trading with Persia, India and China before that, establishing major ports at Hormuz and Muscat. It was Omani seafarers who discovered that monsoon winds could be used to sail their dhows across to the Malabar coast of India and south to Zanzibar, Pemba and the Spice islands of East Africa. Marco Polo, the famous Venetian traveller visited Hormuz in 1271, and in 1498 the Portugese explorer Vasco de Gama sailed from Lisbon to India around the Cape of Good Hope, thus opening the way for future European expeditions to Arabia, India and the Far East. In 1508, tall ships of the Portugese Admiral Alfonso D'Albuquerque's fleet arrived in Muscat. The Portugese sacked and destroyed the town, before continuing along the coast, taking Hormuz seven years later. Muscat was under the control of the Portugese for 144 years, during which time they build the two great forts of Mirani and Jalali, guarding Muscat harbour and the great fort at Muttrah guarding the larger and better anchorage a few kilometres along the rocky coast. By 1590 the powers of Portugal began to wane, and in 1608 the first English ships of Her Majesty Elizabeth 1st East India Company began to establish trading ties all around the Indian Ocean. In 1632 the last Portugese naval commander Ruy Freire de Andrada died in Muscat and a combined English and Persian force soon occupied both Muscat and Hormuz.

Oman has always been a particularly isolated country, surrounded by the great desert of the Rub al-Khali to the west, the inhospitable, dangerous and monsoon-lashed incense coast of south Arabia and the towering ramparts of the Jebel al-Akhdar and the northern mountains inland. Following the demise of Portugal, short periods of domination by Yemenis and Persians followed, but the establishment of the Al Bu Said dynasty of Sultans ruling Muscat and Dhofar, and the rise of religious Imams in the interior made Oman a more divided nation.

Towards the end of the 18th century the rise of the fundamentalist Wahabi Islamic religious sect in the Nejd region of Saudi Arabia began to spread across much of Arabia. Marine trade between Sohar and Muscat with the Gulf was hampered by pirates and raiders from the 'Pirate Coast' of what became the Trucial Oman coast, and is now the United Arab Emirates. Ras al Khaimah was the center of piracy and from here the pirates attacked not only Arab dhows but also ships belonging to the East India Company. In 1809–1810 the English retaliated by attacking the pirate strongholds of Ras al Khaimah and Shinas. Pirates continued to operate and harass the shipping lanes of the Arabian Gulf for many years after. Their story is eloquently told in Sir Charles Belgrave's *The Pirate Coast* (1960).

Bedouin from the tribes living around the margins of the Rub al-Khali regularly traded and raided across the region, but the first crossing of Arabia is credited to Captain George Sadlier who crossed the northern part of the Arabian continent from El Khatif in the Arabian Gulf to Yanbo in the Red Sea in 1819. Sir Richard Burton travelled to Mecca and Medina in 1853, making the first diplomatic contact with the tribes of Saudi Arabia. William Palgrave, an Arabic scholar from Oxford travelled from Syria across the Nejd desert of northern Arabia to Oman in 1862. These exceptional journeys crossed Northern Arabia but did not enter the great dune country of the Rub al-Khali. Charles Doughty travelled widely through Arabia from 1876 to 1878 living with local bedouin tribes. His descriptions in 'Travels in Arabia Deserta' (1888) remain an epic tale of pioneering exploration in Arabia. Captain William Shakespeare made seven separate expeditions into the interior of Arabia between 1910 and 1915 and became a close friend of King Ibn Saud, who was then Emir of Nejd, and later to become the first King of Saudi Arabia.

During the First World War the Arab Revolt against the Ottoman Turks was largely orchestrated by the British with back-room diplomacy from spies like Sir Percy Cox, Gertrude Bell and Harry St. John Philby, and on the ground by forces led by General Allenby and Colonel T. E. Lawrence. Together with the Emir Faisal, Lawrence of Arabia led the Arab revolt against the occupying Turks during the First World War, an expedition that resulted in the epic storming of Aqaba by an army riding camels from the desert, and the subsequent occupation of Damascus in October 1918. Lawrence's book 'Seven Pillars of Wisdom' describes the political and military dramas in epic detail. Britain and France had promised self-determination to the Arabs, but when the post-war political carve-up of the Middle East came, these promises were put on hold, many historians regarding the political outcome as a betrayal of the Arabs. The Sykes-Picot agreement (1916) drawn up by the Allies delineated borders around the British (Palestine, Jordan, Iraq) and French (Syria, Lebanon) mandates for the first time. Following Jewish

Fig. 1.13 Bertram Thomas with Sheik Salih and bedouin in Salalah in 1927, before his first crossing of the Empty Quarter

terrorist atrocities in British-run Palestine, the Balfour Declaration (1917) promised support for a Jewish 'national home' in land that belonged to both settled Palestinians and nomadic Bedouin tribes. This eventually resulted in the formation of the State of Israel on 14th May 1948.

Following the First World War after the defeat of the Turkish Ottoman Empire, France and Britain drew boundaries across northern Arabia that created new countries, mostly with artificial borders. Iraq, Syria, Lebanon, Jordan and Palestine became states, with the borders of Saudi Arabia remaining poorly defined especially in the east. The war affected western and northern Arabia dramatically, but the rest of Arabia including the great Rub al-Khali, the Empty Quarter, and all the interior of Oman and the Trucial coast remained a large blank on the map.

In the winter of 1927–8 Englishman Bertram Thomas who was employed as a *wazir* or advisor to Sultan Said bin Taimur in Muscat, made a 600 mile camel journey through the southern deserts of Arabia to Dhofar, and then in 1929–30 made a further journey 200 miles north of Dhofar along the southern margin of the great sandsea (Fig. 1.13). Bertram Thomas then achieved the first crossing of the great Rub al-Khali desert, the Empty Quarter, in 1930–1 walking and riding with camels from Salalah on the Indian ocean coast to Doha in the Gulf, right across the middle of Arabia. Thomas used bedouin tribesmen to guide him across the uncharted desert. His journey is described in detailed in his classic book '*Arabia Felix, Across the Empty Quarter*' (1932). For this incredible journey he was awarded the Founders Medal of the Royal Geographical Society.

The following year another Englishman, Harry St. John Philby, who was an advisor to King Ibn Saud (Abdul Aziz ibn Abdul Rahman al-Saud) of Saudi Arabia and living in Jeddah, made the second crossing of the Empty Quarter from east to west 1932. He had travelled extensively across Arabia long before his crossing of the Empty Quarter, including an incredible journey by camel from Uqair and Hufuf in eastern Saudi Arabia to Riyadh and Mecca on the Red Sea coast in 1917. Philby was regarded as something of a renegade, advising King Ibn Saud against British interests in Arabia, especially in granting oil concessions. During 1936–7 Philby made several important journeys for Ibn Saud mapping the western tracts of Saudi Arabia from the mountains of Asir and Yemen to the Aden Protectorate and Mukalla in the Hadramaut region of Yemen. His son with his English wife, Kim Philby was the notorious Russian spy who eventually defected to Moscow when his cover was blown. St. John Philby was instrumental in King Ibn Saud granting the Al Hasa oil exploration concession to Standard Oil of California in 1933, and since then, Saudi Arabia has remained allied to the American oil exploration companies.

Probably the greatest of all Western explorers in Arabia, Wilfred Thesiger, born in Ethiopia, bought up in Ethiopia and Kenya and educated in Oxford, made two epic journeys across the Empty Quarter. Despite being denied visas for Saudi Arabia and Oman, Thesiger travelled to Salalah anyway and lived with the Rashid tribe getting completely engrossed into the bedouin way of life. He managed to get a grant from the Locust Research Organisation, and used this as an excuse for his crossing of the Rub al-Khali. His first crossing during 1946–7 was from Salalah on the south coast to the Liwa oasis in Abu Dhabi (Fig. 1.14). Travelling with a few Bait Kathir and Rashid tribesmen, including his companions Salim bin Kabina, Salim bin Ghabaisha, and Muhammed al-Auf and their camels from the southern sands, he walked across the sands of the Rub al-Khali, the heart of Arabia, and explored remote parts of the Empty Quarter (Fig. 1.15). On his return journey, he attempted to explore the great rocky mountains of northern Oman (Fig. 1.16), but was thwarted by the Imam Muhammed al Khalili. When the Imam died in May 1954, Ghalib bin Ali became Imam, and it was he who refused to allow Christians into the Green Mountains, the Jebel al-Akhdar. Instead, Thesiger skirted along the southern flanks of the mountains (Fig. 1.17), and walked across the Hamrat ad-Duru ranges and the Wahiba sands, returning to Dhofar across the Huqf, the home of the legendary unicorn, the Arabian oryx. Thesiger was awarded the

Fig. 1.14 Wilfred Thesiger with his Bait Kathir and Rashid tribesmen, including Muhammed al-Auf, Salim bin Kabina and Salim bin Ghabaisha, Dhofar Copyright Pitt Rivers Museum, University of Oxford (2004.130.6827.1)

Fig. 1.15 Camels climbing a high sand dune during Thesiger's first crossing of the Empty Quarter; photo: Wilfred Thesiger. Copyright Pitt Rivers Museum, University of Oxford (2004.130.22041.1)

Fig. 1.16 Salim bin Kabina and Salim bin Ghabaisha in Oman; photo: Wilfred Thesiger. Copyright Pitt Rivers Museum, University of Oxford (2004.130.22682.1)

Fig. 1.17 Salim bin Kabina and the Oman Mountains, near Jebel Kawr; photo: Wilfred Thesiger. Copyright Pitt Rivers Museum, University of Oxford (2004.130.25553.1)

Founders medal of the Royal Geographical Society in 1948 for his crossing of the Rub al-Khali and exploration of Southern Arabia. His book '*Arabian Sands*' (1959) is the single most iconic and impressive travelogue about Arabian exploration.

It was Wilfred Thesiger who first reported two dome-shaped jebels near Fahud and Natih in January 1947, structures that the oil company explorers were particularly interested in. In February 1948 chief geologist, F. E. Wellings from the Iraq Petroleum Company (IPC, a joint venture with the predecessors of BP, Shell, Total, Exxon/Mobil and Partex), flew over the Oman desert in 1948 and spotted the Fahud domal structure, which then became a prime target for exploration. In 1956 IPC started drilling the first well in Oman on the Fahud dome in the hunt for oil. Following the failure of the first well, Petroleum Development (Oman) comprising 85% Shell 10% Partex and 5% Gulbenkian Foundation took over operations in October 1960 when the rest of the original IPC partners withdrew. On his second crossing of the Empty Quarter during 1947–8, Thesiger explored the remotest regions of the Hadramaut in Yemen and crossed the Rub al-Khali to Abu Dhabi and Buraimi, where he stayed with and became great friends with Sheik Shakbut, paramount Sheik of Abu Dhabi and Sheik Zayed, governor of the eastern province and living in Buraimi, who became the ruler and founder of the UAE. Wilfred Thesiger made many other incredible exploration journeys notably in the marshes of southern Iraq described in his '*Marsh Arabs*' (1964) and in the mountains of Kurdistan, the Hindu Kush and the western Himalaya. His books '*Desert, Mountain and Marsh*' (1979), '*Among the Mountains, Travels in Asia*' (1998), '*A Vanished World*' (2001) and his biography '*A Life of My Choice*' (1987) are epics of exploration travel.

Thesiger continued to travel throughout eastern Arabia, and in 1949 was actually partly financed by IPC, despite his well-known aversion to the oil companies and their plans for oil extraction and development. Thesiger met Sulayman bin Himyar, the paramount sheik of the Bani Riyam tribe and follower of the Imam, who controlled the Jebel al-Akhdar. He asked for British support to recognise his independence from the Sultan of Muscat and Oman. When Thesiger declined, Sulayman bin Himyar refused permission for him to enter the mountains. The Jebel War of 1957–1959 followed when the Sultan's Armed forces together with British mercenaries fought the mountain rebels. This conflict resulted in the routing of the Imam's forces in Jebel Akhdar and the leader of the rebels, Sulayman bin Himyar, together with the Imam Ghalib bin Ali and his brother Talib bin Ali, fled across the border to Saudi Arabia. After this, the tribes in the Jebal al-Akhdar mountains and the interior all swore allegiance to Sultan Said bin Taimur and Oman became united.

Up until the Jebel War the mountains of northern Oman had been completely closed to outsiders. The Bedouin tribes of the interior (Al Bu Shams, Duru and Awamir) rarely ventured into the mountains, and the few hill tribes in the Jebel al-Akhdar were settled in their fortress-like villages perched above great wadis or canyons. These villagers built amazing *falaj* systems, channels bringing water from the *wadis* to irrigate their terraced fields and relied on trading with occasional forays down to the major market towns of Nizwa, Bahla and Ibri south of the mountains and Rustaq to the north, and the coastal ports of Sohar and Muscat further north. Apart from Wellstead who travelled in the 1830s and Percy Cox in the 1920s, the mountains of Oman and UAE were almost completely unknown to outsiders. It was not until the 1920s that the first westerners and geologists like George Martin Lees ventured into the mountains on exploratory treks (Fig. 1.18).

Fig. 1.18 Dawn breaks over the mountains and ridges along Jebel al-Akhdar from the summit of Jebel Shams

---

Bibliography and References

Allfree PS (1967) Warlords of Oman. Robert Hale, London, 191 p
Alston R, Laing S (2012) Unshook till the end of time. Gilgamesh publishing
Belgrave C (1960) The pirate coast. Libraire du Liban, 200 p
Blandford WT (1872) Records of the geological survey of India
Boustead H (1974) The wind of morning. Chatto & Windus, London, 240 p
Carter J (1982) Tribes in Oman. Peninsula Publisher, London
Costa PM (1991) Musandam. Immel Publishing, London, 249 p
Cox P (1925) Some excursions in Oman. Geogr J
Cox P (1929) Across the green mountains of Oman
Curzon HGN (1892) Persia and the Persian question
Dickson HRP (1949) The Arab of the desert. George Allen & Unwin, London, 664 p
Doughty C (1888) Travels in Arabia Deserta. Cambridge University Press
Eccles GJ (1927) The sultanate of Muscat and Oman. J Cent Asiat Soc 14:19–42
Facey W, Grant G (1996) The emirates by the first photographers, Stacey International, 128 p
Glennie K (2005) The desert of Southeast Arabia. GeoArabia, Bahrain
Henderson E (1988) This strange eventful history, Motivate, 242 p
Lawrence TE. Seven pillars of wisdom
Lees GM (1928) The physical geography of South-Eastern Arabia. Geogr J 71:441–466

Miles S (1884) J Excursion Oman, in Southeast Arabia

Miles S (1919) The countries and tribes of the Persian Gulf

Miles S (1920) On the border of the Great Desert: A journey in Oman

Monroe E (1973) Philby of Arabia, Faber & Faber, London, 334 p

Morris J (1957) Sultan in Oman, Faber & Faber, London, 140 p

Morton M (2006) In the heart of the desert. Green Mountain Press, 282 p

Morton M (2013) Buraimi: the struggle for power, influence and oil in Arabia, I.B.Taurus, London, 286 p

Palgrave WG (1865) Narrative of a year's journey through Central and Eastern Arabia

Paxton J. History of PDO. Middle East Archive Center, St-Anthony's College, Oxford (unpublished)

Petroleum Development Oman (2009) Oman faces and places, Muscat, Oman

Peyton WD (1983) Old Oman. Stacey International, London, 128 p

Phillips W (1967) Oman—a short history. Longmans, London

Searle P (1979) Dawn over Oman. George, Allen & Unwin, 146 p

Searle M (2013) Colliding continents. Oxford University Press, 438 p

Sheridan D (2000) Fahud, the Leopard Mountain. Vico Press, Dublin, 268 p

Skeet I (1974) Muscat and Oman, the end of an era. Faber and Faber, London, 224 p

Stanton-Hope WE (1951) Arabian adventurer

Stark F (1936) The Southern Gates of Arabia. John Murray, London, 327 p

Stiffe A (1897) Capt. Ancient trading centres of the Persian Gulf. Geogr J

Thesiger W (1948) Across the Empty Quarter. Geogr J 111:1–21

Thesiger W (1959) Arabian sands. Longmans, 326 p

Thesiger W (1964) Marsh Arabs. Longmans, Green & co., 233 p

Thesiger W (1979) Desert, marsh and mountain. Collins, 304 p

Thesiger W (1987) The life of my choice. Harper Collins

Thesiger W (1987) Visions of a Nomad. Harper Collins, 224 p

Thesiger W (1998) Among the mountains, travels through Asia. Harper Collins, 250 p

Thesiger W (1999) Crossing the sands. Motivate, 176 p

Thesiger W (2001) A vanished world

Thomas B (1931) Alarms and excursions in Arabia. George Allen & Unwin, London, 316 p

Thomas B (1932) Arabia felix, across the empty quarter of Arabia. Jonathan Cape, London, p 397

Villiers AT (1952) The Indian Ocean

Ward P (1987) Travels in Oman. The Oleander Press, Cambridge, England, 572 p

Wellstead JR (1837) Narrative of a journey into the interior of Oman. J R Geogr Soc, London

Wellstead JR (1838) Travels in Arabia, vol 1

Wilkinson J (1987) The imamate tradition of Oman. Oxford University Press

Wilkinson JC (1991) Arabia's frontiers: the story of Britain's boundary drawing in the desert. I.B. Taurus, London, 400 p

Winser N (1989) The sea of sands and mists. Century, London, 199 p

Exploration of the Oman–UAE Mountains

Despite being located along the eastern margin of Arabia where the Indian Ocean passes into the Gulf of Oman and the Arabian Gulf, Oman was one of the most remote and inaccessible parts of Arabia before the rush of development following the discovery of oil. A mountain chain more than 700 km long stretches from the Musandam peninsula and the Straits of Hormuz in the north to the eastern tip of Arabia near Ras al-Hadd. A narrow coastal plain, the fertile Batinah coast separates the mountains from the Gulf in northern Oman, and east of Muscat the mountains descend directly into the ocean. Oman has a rich sea-faring tradition because it is located in a critical position along the East African trading routes from the spice islands of Zanzibar and Pemba, the rich Malabar coast of India and the barren Makran coast of southern Baluchistan. It also controls the gateway to the Arabian (Persian) Gulf through the Straits of Hormuz between Musandam and the Iranian coast to the north. The seas off Oman teem with fish, attracted by the upwelling of plankton-rich currents along the western side of the Indian Ocean. Omani traders used the monsoon winds to sail large ocean-going dhows south to East Africa using the southwest directed winter winds and Somali current, returning along the western Malabar coast of India with the northeast blowing summer monsoon winds, trading with Africa, India and all points east to the Indonesian islands and beyond.

The Portuguese were the first western visitors to Oman with ships calling into re-supply with fresh water and food on their way east to the spice islands of the East Indies. The two great forts guarding Muscat harbour, Jalali and Mirani, were built by the Portuguese and completed in 1581, but the Portuguese were expelled from Muscat soon after, in 1624, by Sultan bin Saif who founded the Ya'ruba dynasty and started negotiations with the British East India company. He ruled from his newly built forts at Rustaq and Nizwa. Although Forts Jalali and Mirani in

© Springer Nature Switzerland AG 2019
M. Searle, *Geology of the Oman Mountains, Eastern Arabia*,
GeoGuide, https://doi.org/10.1007/978-3-030-18453-7_2

Fig. 2.1 **a** Forts at entrance to Muscat harbour, **b** Fort Jalali, Muscat harbour, **c** Fort Mirani, Muscat harbour, **d** Jalali from Muscat 1968, **e** Nakhl fort, **f** Muscat from Fort Mirani

Muscat and the Mattrah fort were built by the Portuguese all other forts throughout Oman were built by Omani craftsmen (Fig. 2.1). Many of these spectacular castles still stand today (Fig. 2.2), some of them recently restored to their former glory. The most impressive mountain forts include those at Jabreen, Nizwa, Bahla, Ibri, al Hazm and Rustaq.

Fig. 2.2 a, **b** Rustaq fort, 1970 **c** Sulayf fort along the western edge of the mountains, **d** Misfah village on southern slopes of Jebel Akhdar

The British East India company needed supply ports to service their newly flourishing trade with India and opened a consulate (later becoming an Embassy) in Muscat, adjacent to the Royal Palace on Muscat beach and described by Lord Curzon, the future Viceroy of India, as the 'handsomest building in Muscat'. Many early travellers made brief journeys inland from Muscat and the earliest maps come from these intrepid explorers, notably Capt. Arthur Stiffe who surveyed the Muscat area in 1860, and drew some wonderful drawings of the jagged mountain coastline around Muscat ('*Ancient Trading Centres of the Persian Gulf*' Geographical Journal, 1897). Lieut. Wellstead visited Nizwa in 1835 and noted that the prosperous town had a thriving industry making copper coffee pots and goldsmiths and silversmiths making the famous Oman *khanjahs* (traditional Omani curved daggers) and jewellery. Copper was sourced from the old mines in Wadi Jizzi, where smelting is known to date back to the Early Bronze age (third millennium BC). Wellstead wrote about these early forays into the interior of Oman in his '*Travels in Arabia*'

(1838). Several of the earliest British geologists and naturalists also stopped off in Muscat on their way out India. Amongst these, W. T. Blandford noted from his visit in 1872 that the rocks around Muscat were unusual, being composed of dark-coloured serpentine, the first recognition that the ophiolitic rocks in Oman were distinctly different from the usual continental sedimentary and granitic rocks.

Samuel Miles made several incredible journeys by dhow and on foot and by camel across the Oman mountains including a dhow journey from Muscat east to Sur and Al Ashkara in 1874, another from Sohar to Buraimi and Abu Dhabi in 1875 described in '*The Countries and Tribes of the Persian Gulf*' (1919), and from Muscat to Jabreen south of the Jebel al-Akhdar in 1885. On this trip Miles saw the ancient mud-brick towns of Nizwa and Bahla with their splendid large castle forts and intricate bustling souqs with narrow alleyways surrounded by lush green lime and date groves. He also saw the towering limestone mountains of Jebel Misht and Jebel Misfah, two of the most impressive peaks in Oman. Perhaps his most remarkable journey was the one in 1884 from Muscat and Mattrah south to Wadi Tayyin though the Wadi Daykah gorge (the 'Devils Gap') and then east and north, returning to the coast at Quriat. How they managed to get fully laden camels through the precipitous gorge of Wadi Daykah is wonderfully described in his '*Journal of an Excursion in Oman, Southeast Arabia*' (1884).

Sir Percy Cox was the Political Agent and Honorary Consul in Muscat from 1899–1904 and also travelled extensively across Oman. He was the first European to travel between Ras al Khaimah on the Trucial coast across the desert and western mountains to Dhank. He trekked from Abu Dhabi to Buraimi and explored along the mountain ranges of Jebel Akhdar and Jebel Nakhl, climbing up to the Saiq plateau and the village of Bani Habib ('*Across the Green Mountains of Oman*' and '*Some Excursions in Oman*', Geographical Journal, 1925). Cox was also the first European to explore the remote and mountainous Musandam peninsula, voyages that had to be made by *dhow* as much of Musandam is only accessible from the sea. He described the Elphinstone inlet (Khor ash Sham, east of Khasab) on the west side of Musandam and the Malcolm inlet (Khor Habalayn)—a deep fjord carved into sheer limestone cliffs on the east coast and, in those days, almost completely inaccessible from the land.

2.1   Earliest Geological Explorations

Henry Pilgrim was the first to attempt to make a geological map of Oman following his travels in 1904–5, but the first proper geological expedition to the northern mountains of Oman was commissioned by the D'Arcy Exploration company (a subsidiary of Anglo-Persian Oil Company) in 1925. The party included two geologists, G. M. Lees and K. W. Gray, a political officer, J. G. Eccles, a naturalist Joseph Fernandez, and an Arabist A. F. Williamson, known as 'Haji Abdulla'. Williamson had converted to Islam and spent more than 25 years exploring remote parts of Yemen and western Arabia, acquiring an unprecedented knowledge of the bedouin way of life. During the war he was attached to the intelligence branch in Iraq. His life story has been told in W. E. Stanton-Hope's biography '*Arabian Adventurer*' (1951). The main aim of the D'Arcy expedition was geological, but they were unable to access the high Jebel al-Akhdar range which was under the control of the Imam and Sheikh Isa bin Salih. Instead, they trekked along Wadi Hawasina and explored the mountains to the north around Wadi Bani Umar and Wadi Ahin (Haybi). They also travelled north to Shinas turning into the mountains along Wadi Hatta and returning to Sohar along Wadi Jizzi. '*Mapping in this igneous country is very difficult. Many of the peaks are unscaleable owing to the steepness of the slope and the loose rubble, and if by some good fortune one does reach a view point the surroundings consist of a maze of jagged peaks among which it is very difficult to select one which will be recognizable from a different angle*' (Lees 1928).

The D'Arcy expedition travelled mainly though country formed of the ophiolite and the Hawasina complex deep-sea sedimentary rocks, and it was G. M. Lees who first proposed that the serpentinites and the underlying Hawasina series cherts were exotic, emplaced into their present position by thrusting (Fig. 2.3). In the Wadi Hawasina and Wadi Bani Umar he saw the mixture of '*diorite, gabbro and green serpentinite, with lava flows ranging from keratophyres to basalts*'. These green serpentinites, red cherts and black basalts that make up the classic 'coloured mélanges' typical of Tethyan ophiolites. Lees emphasised the '*Cretaceous movements were deep-seated and has produced at the surface a complex of great overthrust sheets or nappes, rather than simple folding*'. Lees had the advantage of having read Wegener's (1922) hypothesis of continental drift '*elaborated and applied in a most masterful manner to the tectonics of Asia*' by Argand (1924). Wegener had proposed that the Indian continent was initially attached to Madagascar and South Africa before separating and drifting north to collide with Asia. This was an astounding foresight considering that it was 45 years before the plate tectonic revolution of the late 1960s and 1970s. Argand proposed that India had underthrust Asia uplifting the Himalaya and Tibetan Plateau in another brilliant work of the same time. Lees was almost certainly influenced by both Wegener and Argand although he did emphasise, confusingly, that '*I am not a pro-Wegenerist*'. His paper '*The Geology and Tectonics of Oman and parts of southeast Arabia*' (1928) was, however, the first interpretation for the Cretaceous thrust origin of the

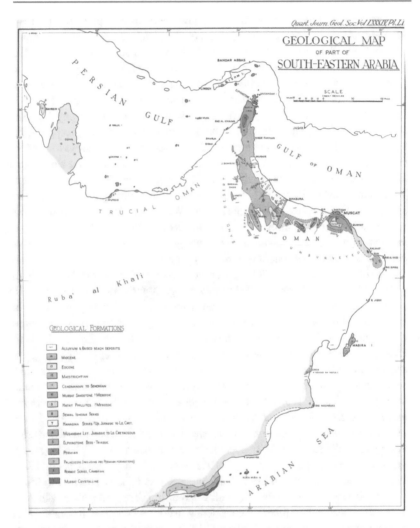

Fig. 2.3 The first geological map of part of northern Oman and UAE, published by G. M. Lees in 1928

Oman serpentinites and the Hawasina cherts. Lees also correlated the Oman Mountains westwards to the 'Persian arc' and the Alps, and eastwards to the mountains of western India.

2.2 Discovery of Oil

Sultan Said bin Taimur had granted the concession for oil exploration to Petroleum Development (Oman and Dhofar) in 1937. The Dhofar part was dropped when that area was abandoned in 1951 and the company became known as Petroleum Development (Oman) Ltd or PDO. Ever since Thesigers' excursions in interior Oman, exploration geologists from PDO had long been interested in Jebel Fahud, a classic anticlinal structure typical of oil-bearing structures in Iran and Iraq. The core of the anticline has been eroded to reveal the older underlying rocks and it was a tempting target for drilling to the reservoir rocks at depth in the middle of the fold. Geologists had first seen the structure and its northern neighbour Jebel Natih in the late 1940s from the air during flights between the RAF bases at Sharjah and Masirah Island. Chief Geologist F. E. Wellings identified several great fold structures in the white Cenozoic limestones along the southwestern flank of the Oman Mountains, including the Jebel Hafit anticline near Buraimi and those at Fahud and Natih. These areas soon became primary targets for more detailed geological exploration. It was already known from wells drilled across the border in Saudi Arabia that the Cretaceous Wasia Formation was the prime oil reservoir with the overlying shales of the Late Cretaceous Aruma Group acting as the seal.

The problem was that successive attempts to survey both the Buraimi and Fahud areas ran up against tribal uprisings and the religious fervour of the Imam, Ghalib bin Ali, who together with his brother Talib bin Ali, seemed to control much of the interior. The company realised that Arabists and politicians were needed to negotiate with the fractious sheiks, who all wanted control of the oil rights. The company hired a new liaison officer, Edward Henderson who arrived on the scene in 1948, and was based initially in Dubai. Both Henderson and the company's representative, Dick Bird had built strong bonds with the tribes and especially with Sheik Zayed of Abu Dhabi, the same Sheik Zayed who hosted Wilfred Thesiger in Al Ain, the spring-fed oasis beneath Jebel Hafit, and adjacent to the Omani village of Buraimi. Wilfred Thesiger and Sheik Zayed used to hunt the great houbara bustards from camels using hunting falcons (*Shaheen*) and salukis, and both men were deeply impressed with the simple life of the bedouin.

It was largely through the political skills of Dick Bird and Edward Henderson that the tribes of the Dhahira desert from Buraimi south to Fahud, including the Duru, became united behind the Sultan Said bin Taimur. A major international dispute arose in 1952 when a Saudi force under Turki bin Utayshn occupied an

Omani village, Hemasa, in the Buraimi oasis. The Saudis had long cast covetous eyes on the potential oil-bearing fold structures of Jebel Hafit, and even had plans to take over parts of Abu Dhabi, but this was an attempt to divide the followers of the Sultan and the Imam and a clear attempt to annexe a potentially oil-rich part of the Oman and UAE desert. The US government supported ARAMCO and the Saudi claim whilst the British strongly backed the Sultan. The stalemate ended in 1955 when a detachment of the Trucial Oman Levies forced the Saudis to leave and occupied the fort at Buraimi. Sheik Saqr of Buraimi and other renegade sheiks supporting the Imam were driven out into the desert. The whole Buraimi dispute is described in Mike Morton's book '*Buraimi: the Struggle for Power, Influence and Oil in Arabia*' (2013).

Fahud was in the middle of territory under the control of the notoriously garrulous Duru tribe who were flipping between giving support to the Imam, Ghalib bin Ali al-Hinai and the Sultan, Said bin Taimur. In 1954 the Sultan's armed forces (Muscat and Oman Field Force) laid siege to the Imam's supporters holed up in the fort at Ibri, and it was only the negotiating skills of Edward Henderson, the British political representative that avoided bloodshed. Shortly after the Buraimi dispute, lingering rebellions from tribes loyal to the Imam continued until the Sultan requested assistance from the British to defeat the rebels. The Jebel War of 1957–1959 followed when the Sultan's Armed forces together with British mercenaries defeated the mountain rebels, stormed the Saiq plateau and bombed the fort of Sulayman bin Himyar at Tanuf. Sulayman, together with the Iman Ghalib bin Ali, fled across the border to Saudi Arabia and the mountain villages of Jebel al-Akhdar were finally brought into the fold of the Sultan.

As soon as the Imam's supporters had been defeated, geological mapping could begin. Over four weeks during November–December 1954, the geologists, Mike Morton, Don Sheridan, Rodney Collomb, Nick Fallon and Jim McGinty mapped the Fahud, Natih and Maradi regions and the company decided that drilling would go ahead. An airstrip was built so that Dakotas could supply the camp with food, water and supplies. Land Rover tracks emanated out from Fahud as geologists explored south towards the Umm al Sammim quicksands bordering the Empty Quarter and east towards the great Wahiba sandsea. An excellent description of these pioneering times in the Oman desert is given in Mike Morton's book '*In the Heart of the Desert*' (2006).

On 18th January 1956 the first oil well Fahud-1 was spudded and drilling began. Although there were some minor seeps of oil and gas, the well was dry and after drilling 12,235 feet it was abandoned. During the geological mapping and early seismic studies, a large fault was discovered crossing the Fahud anticline and the geologists requested another well be drilled across the fault. However, the company management was reluctant and refused. As a result, three of the original IPC partners decided to withdraw leaving Shell with 85% stake-holding, and Partex (15%) to continue exploration. A major new drilling programme was initiated in 1962 and large discoveries were soon made in similar anticlinal structures at Yibal, west of Fahud, and Natih, east of Fahud. A second well was drilled on the original Fahud anticline in 1964, Fahud-2 across the fault from the first hole, and this time struck oil in considerable quantity. The Fahud-2 well stopped production only in 2005, 41 years after the first discovery. Later seismic surveys revealed that the fault at Fahud dropped the oil-bearing Wasia limestones down by nearly 3000 feet and the Fahud-1 well missed the payzone by only a few hundred meters. Julian Paxton described Fahud-1 as 'the unluckiest oil well ever drilled'.

With the accession of Sultan Qaboos bin Said in July 1970 (Figs. 2.4 and 2.5), it became imperative to discover more oil reserves so that Oman could develop as other Gulf countries were doing. A major exploration programme was launched starting with geological mapping of the entire Oman Mountains, the geological key to interpret the structure and stratigraphy of the foreland. PDO had a small camp on

Fig. 2.4 Sultan Qaboos bin Said arriving in Bait al Falaj airstrip, Muscat, July 1970

Fig. 2.5 Sultan Qaboos bin Said visiting Petroleum Development Oman main office with MD Francis Hughes and Geoff Searle, August 1970; photos courtesy of Petroleum Development Oman Ltd.

Fig. 2.6 Ras al Hamra, 1970, with Fahal Island in distance

the headland at Ras al Hamra (Fig. 2.6) and staff and supplies were flown in by Dakotas to a newly constructed airstrip at Azaiba (Fig. 2.7). The Batinah coast was totally undeveloped and the sweeping stretch of white sand beach extended from Qurum (Fig. 2.8), all the way up the Batinah coast to Sohar and beyond. In the

Fig. 2.7 Dakota aircraft at Azaiba airport, gateway to Oman, 1968

Fig. 2.8 Qurum beach, 1969

Fig. 2.9 Izki oasis, Oman Mountains, 1977

interior large villages like Izki (Fig. 2.9) were all made of mud brick and palm fronds. There were no tarmac roads anywhere in the country and access into the mountains was entirely by 4-wheel drive Land Rovers (Fig. 2.10). Some of the villages in the mountains were extremely remote and the only access was walking along the large wadi systems and climbing (Figs. 2.11 and 2.12).

2.3 Geological Mapping

Following the discovery of significant quantities of oil and gas at Yibal, Natih and Fahud, PDO decided that it was of crucial importance to find out what made up the geology of the northern mountains. The strata in the Jebel Akhdar range dipped south beneath the ophiolitic rocks around Nizwa and Bahla and was continuous with sub-surface strata that extended south beneath the desert to Fahud, Natih and the southern mountain structures. The thrusted origin of the Hawasina series and Semail ophiolite, as first proposed by Lees, was challenged by the subsequent studies of Morton (1959), Tshopp (1967) and Wilson (1969) who all proposed an auto-chthonous origin of the Hawasina series lying above the Permian to Cenomanian

Fig. 2.10 Land Rover in Wadi Sarami, Haybi district

Fig. 2.11 Wadi Sarami access to Haybi corridor

Fig. 2.12 Wajma village, beneath the north face limestone cliffs of Jebel Shams

limestones of Jebel Akhdar. Wilson (1969) wrote '*it is concluded from field data that the Hawasina was deposited in its present location during latest Cretaceous time*', and '*the flood eruption of Semail ultrabasic magma which blanketed the abyssal landscape and cooled slowly under great hydrostatic pressure along the axis of the trough. Compressional features in Hawasina are attributed to contemporaneous gravity slumping of the beds on the steep northeast limb of the trough.*' Wilsons' paper is an excellent summary of the stratigraphy of the shelf carbonates as known at that time, but his interpretation of the structure and origin of the Hawasina deep-sea sedimentary rocks and the Semail ophiolite in the mountains is at odds with all subsequent work.

During two winter campaigns in 1966–1968 Shell geologists set about a major geological study along the entire length of the Oman Mountains, using field bases at Azaiba and Ibri, with Land Rovers and helicopter support. Led by Ken Glennie, the team included Michel Boeuf, W. 'Pit' Pilaar, Mike Hughes-Clarke and Mark Moody-Stuart who were responsible for all the sedimentary units, and Ben Reinhardt who was responsible for mapping and interpreting the Semail ophiolite. They also had the advantage of several crucial unpublished PDO reports from geologists such as Jean Haremboure and Wim Horstink (1967) who proposed a Mesozoic nappe hypothesis, as opposed to Morton and Wilson's authochthonous model for the structure of the Hawasina and Semail ophiolite. The team had complete aerial photographic coverage of Oman and the UAE, collected more than 2,500 samples and made over 12,000 thin sections for microscope studies. Glennie's team produced an excellent and comprehensive memoir (Glennie et al. 1974) that included two map sheets of the geology of the Oman Mountains at 1:500,000 scale, structural cross-sections, detailed stratigraphic and palaeontological log sections of every formation, a volume of field photographs and locality illustrations, and two volumes of a geological report. This comprehensive work provided by far the most detailed and precise record of the geology of the Oman mountains, and remains the primary source of information to this day. A summary paper was published in the Bulletin of the American Association of Petroleum Geologists in 1973. Reinhardt (1969) also wrote a separate paper on the genesis and emplacement of the ophiolite, the first real detailed study of the structure and petrology of the ophiolite. Reinhardt mistakenly thought that the gabbro-basalt suite ranged in age from mid-Permian to Cretaceous, similar to the age of the autochthonous shelf carbonates with the ultramafic rocks 'pre-geosynclinal'. We now know from U–Pb dating of zircons extracted from the plagiogranites and gabbros that the entire crustal section of the Oman ophiolite formed during the Cenomanian-Turonian (Late Cretaceous) time around 96–94 million years ago. Reinhardt also made the first observations of the 'Metamorphic sheet' along the base of the ophiolite and proposed a model similar to that of formation of the ophiolite along a modern mid-ocean ridge.

Ken Glennie's team dated the authochtonous shelf carbonate succession using microfossils, as well as the allochthonous pelagic sediments, using radiolaria biostratigraphy. They found radiolarian fossils in cherts interbedded with and on top of the pillow lavas that were of Cenomanian to early Senonian (Late Cretaceous). They concluded that the Hawasina deep-sea sedimentary rocks were deposited during the Permian to Cenomanian time northeast of the present-day mountains, and that the rocks of the Sumeini Group (shelf margin), Hawasina complex (Tethyan ocean basin) and Semail ophiolite were emplaced tectonically into their present position from NE to SW along the Arabian continental margin. Crucially the Glennie team found that many of the 'Oman Exotics' immediately beneath the ophiolite were thrust onto deep-water pelagic cherts of the Hawasina complex. After emplacement of the Semail ophiolite and underlying thrusts, the whole allochthon was covered unconformably by shallow marine Maastrichtian to early Tertiary (Cenozoic) limestones. The mountains were then uplifted during the Oligocene–Miocene. This interpretation has been backed up and largely accepted by almost all subsequent workers.

The Glennie et al. (1974) memoir and geological map laid the foundations for all future studies in the Oman and UAE mountains. Oman remained a closed country up until the mid-1970s with only PDO employees and their families allowed into the country. With the accession of Sultan Qaboos in July 1970, the country began to open up to the outside world and embarked on an extremely rapid transition with new developments springing up across the country. In 1974–5 two groups were given permission to carry out geological field-based research in the mountains. Bob Coleman and Cliff Hopson from the US Geological Survey and the University of California, both ophiolite experts from California, were granted permission to survey a strip across the mountains from Muscat south to Ibra. Over the next six winter seasons they, with several Ph.D. students, mapped this transect in detail and carried out comprehensive geochemical and isotopic studies on the ophiolitic rocks. At the same time a group from the Open University (OU) in the UK, headed by Ian Gass, with John Smewing, Adrian Lewis and Steve Lippard in charge of the field parties was given permission to work in the northern Oman mountains. Ian and John had spent several years studying the Troodos ophiolite in Cyprus and their background knowledge was invaluable to the success of this group. The OU group, including seven Ph.D. students over 10 years, based in Sohar, mapped the entire ophiolite in Northern Oman, from the UAE border south to Rustaq, and published a series of geological maps at a scale of 1:100,000.

I was lucky enough to be one of the Ph.D. students on Ian Gass' Oman project, starting in the winter of 1976–7. I was given a completely free range with my project and after two weeks of driving all over ophiolite of the northern Oman mountains I decided to work on the Metamorphic sole along the base of the ophiolite and the thrust structures beneath. I mapped four areas in detail, the Sumeini and Asjudi (Jebel Qumayrah) Windows on the western flank of the ophiolite, the Haybi–Wadi Ahin–Wadi Hawasina area in the middle of the ophiolite, all in Oman, and a reconnaissance map of the Dibba zone in the United Arab Emirates. It was an exciting time to be working on ophiolites. The plate tectonic revolution of the 1960s and 1970s had established that sea-floor spreading resulted from mantle convection, and ocean crust was formed along mid-ocean ridges such as the Mid-Atlantic ridge and the East Pacific Rise. Subduction zones were characterised by deep-water trench sediments, high-pressure blueschists and eclogites and serpentinite mélanges. Techniques of constructing balanced cross-sections had been proposed, and it was possible to restore these sections to obtain minimum amounts of crustal shortening across fold and thrust belts. Geochemical techniques, particularly obtaining major, trace and rare-earth element analyses of basaltic rocks enabled geologists to compare ancient lavas with modern tectonic settings. Polished thin-sections of rock samples could be analysed by microprobe to obtain precise chemical compositions of tiny mineral grains in metamorphic and igneous rocks. The science of thermobarometry, obtaining pressures and temperatures of metamorphism using mineral pairs like garnet and pyroxene, was then in its infancy but provided an exciting additional dimension to the whole story.

In those days there were no tarmac or black-top roads anywhere in Oman and off-road 4-wheel drive Land Rover access was the only way to get into the field (Fig. 2.11). We used to load the Land Rover up with food and water supplies for 10 days, a rack of jerry cans full of petrol on the roof, spades and ropes for digging the vehicle out of sand or wadi gravels, and head off into the mountains with our maps, compasses and hammers. It was a wonderful life, camping out under the stars, hiking over the mountains and along the wadis all day, and relaxing by a camp-fire in the evenings. Gradually my maps were filled in and cross-sections constructed, and then we would all meet back at our house in Sohar and discuss our findings. Sohar was then a quiet sleepy fishing town on a wide sandy beach with an old mud-brick fort, a quiet fruit and vegetable souq and only a few Land Rovers on the sandy unpaved roads. Every morning the *shasha* reed boats, *badans* and small dhows used to land their boats on the beach and the fishermen would sell off their catch to villagers. Once a month we used to drive back to Muscat to connect with civilisation and talk to our sponsors in the Ministry and to PDO friends.

During the 1980s and 1990s several other groups started work on the Oman ophiolite. Of particular note was Tjerk Peters and his group from the University of Berne, Switzerland who had previously mapped parts of the UAE section and later mapped the Masirah and Batain coast ophiolites. French groups led by Thierry Juteau and René Maury from the University Bretagne Occidentale, and Adolfe Nicolas and Francoise Boudier from the University of Montpellier in France who, with several colleagues and Ph.D. students, spent more than 30 years studying the ophiolite along the length of the mountains. The Bureau de Recherches Géologiques et Minieres (BRGM) also spent more than 20 years mapping the entire Oman part of the mountains and published a whole series of detailed geological maps at scales of 1:100,000 and 1:250,000. Several British research groups worked across the Oman Mountains during the period 1980–2010. Many results were published in two volumes of the Geological Society of London, Special Publications volume 49, *The Geology and Tectonics of the Oman region* (1990; edited by A. H. F. Robertson, M. P. Searle and A. C. Ries) and volume 392, *Tectonic Evolution of the Oman Mountains* (2014; edited by H. R. Rollinson, M. P. Searle, I. A. Abbasi, A. Al-Lazki and M. H. Al-Kindy). The British Geological Survey spent 7 years mapping the entire UAE part of the mountains and published a series of geological maps at scales of 1:50,000 and 1:100,000. Fugro and Western Geophysical conducted geophysical surveys (regional gravity and magnetics, and seismic) and shot four deep seismic lines across the UAE. Mike Searle from the University of Oxford in UK continued geological projects with numerous Ph.D. (D.Phil), and MSc students, both in Oman and UAE continuously from 1990 to the present day. Mohammed Ali from Oxford University and the Petroleum Institute, Abu Dhabi, also led a 15-year project combining geophysics with onland geological studies in the Musandam peninsula and UAE part of the mountains, resulting in several key publications. Many others continued field-based studies on the ophiolite and metamorphic sole, as well as the sedimentology and structure of the Haybi, Hawasina and Sumeini complexes. These workers will be referred to in more detail in the following chapters.

Bibliography and References

Allfree PS (1967) Warlords of Oman. Robert Hale, London, 191 p
Alston R, Laing S (2012) Unshook till the end of time. Gilgamesh publishing
Argand E (1924) Le tectonique de l'Asie. Proceedings of 13th International Geological Congress
Belgrave C (1960) The pirate coast. Libraire du Liban, 200 p

Blandford WT (1872) Records of the geological survey of India

Boustead H (1974) The wind of morning. Chatto & Windus, London, 240 p

Carter J (1982) Tribes in Oman. Peninsula Publisher, London

Costa PM (1991) Musandam. Immel Publishing, London, 249 p

Cox P (1925) Some excursions in Oman. Geogr J

Cox P (1929) Across the green mountains of Oman

Curzon HGN (1892) Persia and the Persian question

Dickson HRP (1949) The Arab of the desert. George Allen & Unwin, London, 664 p

Doughty C (1888) Travels in Arabia Deserta. Cambridge University Press

Eccles GJ (1927) The sultanate of Muscat and Oman. J Cent Asiat Soc 14:19–42

Facey W, Grant G (1996) The emirates by the first photographers, Stacey International, 128 p

Glennie K (2005) The desert of Southeast Arabia. GeoArabia, Bahrain

Glennie KW, Boeuf MG, Hughes-Clarke MHW, Moody-Stuart M, Pilaar WF, Reinhardt BM (1973) Late Cretaceous nappes in the Oman mountains and their geologic evolution. Bullet Am Assoc Petrol Geol 57:5–27

Glennie KW, Boeuf MG, Hughes-Clarke MHW, Moody-Stuart M, Pilaar WF, Reinhardt BM (1974) Geology of the Oman mountains. Verhandelingen Koninklijk Nederlands geologisch mijnbouwkundidg Genootschap 31:423

Haremboure J, Horstink J (1967) Unpublished PDO Report

Henderson E (1988) This strange eventful history, Motivate, 242 p

Lawrence TE. Seven pillars of wisdom

Lees GM (1928) The physical geography of South-Eastern Arabia. Geogr J 71:441–466

Miles S (1884) J Excursion Oman, in Southeast Arabia

Miles S (1919) The countries and tribes of the Persian Gulf

Miles S (1920) On the border of the Great Desert: A journey in Oman

Monroe E (1973) Philby of Arabia, Faber & Faber, London, 334 p

Morris J (1957) Sultan in Oman, Faber & Faber, London, 140 p

Morton DM (1959) The geology of Oman. 5th World Petroleum Congress, New York

Morton M (2006) In the heart of the desert. Green Mountain Press, 282 p

Morton M (2013) Buraimi: the struggle for power, influence and oil in Arabia, I.B.Taurus, London, 286 p

Palgrave WG (1865) Narrative of a year's journey through Central and Eastern Arabia

Paxton J. History of PDO. Middle East Archive Center, St-Anthony's College, Oxford (unpublished)

Petroleum Development Oman (2009) Oman faces and places, Muscat, Oman

Peyton WD (1983) Old Oman. Stacey International, London, 128 p

Phillips W (1967) Oman—a short history. Longmans, London

Reinhardt BM (1969) On the genesis and emplacement of ophiolites in the Oman mountains geosyncline. Schweiz Minerl Petrogr Mitt 49:1–30

Searle P (1979) Dawn over Oman. George, Allen & Unwin, 146 p

Searle M (2013) Colliding continents. Oxford University Press, 438 p

Sheridan D (2000) Fahud, the Leopard Mountain. Vico Press, Dublin, 268 p

Skeet I (1974) Muscat and Oman, the end of an era. Faber and Faber, London, 224 p

Stanton-Hope WE (1951) Arabian adventurer

Stark F (1936) The Southern Gates of Arabia. John Murray, London, 327 p

Stiffe A (1897) Capt. Ancient trading centres of the Persian Gulf. Geogr J

Thesiger W (1948) Across the Empty Quarter. Geogr J 111:1–21

Thesiger W (1959) Arabian sands. Longmans, 326 p

Thesiger W (1964). Marsh Arabs. Longmans, Green & co., 233 p

Thesiger W (1979) Desert, marsh and mountain. Collins, 304 p

Thesiger W (1987) The life of my choice. Harper Collins

Thesiger W (1987) Visions of a Nomad. Harper Collins, 224 p

Thesiger W (1998) Among the mountains, travels through Asia. Harper Collins, 250 p

Thesiger W (1999) Crossing the sands. Motivate, 176 p

Thesiger W (2001) A vanished world

Thomas B (1931) Alarms and excursions in Arabia. George Allen & Unwin, London, 316 p

Thomas B (1932) Arabia felix, across the empty quarter of Arabia. Jonathan Cape, London, 397 p

Tshopp RH (1967) The general geology of Oman. 7th World Petroleum Congress, Mexico

Villiers AT (1952) The Indian Ocean

Ward P (1987) Travels in Oman. The Oleander Press, Cambridge, England, 572 p

Wellstead JR (1837) Narrative of a journey into the interior of Oman. J R Geogr Soc, London

Wellstead JR (1838) Travels in Arabia, vol 1

Wilkinson J (1987) The imamate tradition of Oman. Oxford University Press

Wilkinson JC (1991) Arabia's frontiers: the story of Britain's boundary drawing in the desert. I.B. Taurus, London, 400 p

Wilson HH (1969) Late Cretaceous eugeosynclinal sedimentation, gravity tectonics, and ophiolite emplacement in Oman mountains. Am Assoc Petrol Geol 53:626–671

Winser N (1989) The sea of sands and mists. Century, London, 199 p

Part II
Geology

Geology of the Oman–United Arab Emirates Mountains

3

The geological record in the Oman–UAE mountains and the interior from Precambrian to the present day can be broadly divided into seven major tectonic units, based on outcrop mapping, well data, magnetic and gravity data (Glennie et al. 1973, 1974; Hughes-Clark 1990; Robertson and Searle 1990; Sharland et al. 2004). From oldest to youngest these include (1) Precambrian basement, (2) Neoproterozoic sedimentary sequences and salt basins, (3) Palaeozoic sequences, (4) Middle Permian to Cenomanian shelf carbonate sequence, and its outboard equivalent (5) the Tethyan oceanic sediments of the Hawasina and Haybi complexes, (6) Aruma flexural foreland basin, and (7) the post-obduction Maastrichtian and Cenozoic sedimentary cover rocks (Fig. 3.1).

3.1 Precambrian Basement Rocks

Precambrian metamorphic basement outcrops in Saudi Arabia and interior Yemen are mainly composed of Pan-African metamorphic rocks formed in island arcs, suture zones and subduction-accretion complexes. Older Archean and Palaeoproterozoic rocks are found in Yemen, but most of the Arabian Shield is composed of overlying Proterozoic rocks. The term 'Pan-African' event includes all Neoproterozoic ($\sim$800–520 million years) deformation and magmatism associated with the assembly of the Gondwana supercontinent. A prominent NW-SE trending sinistral transpression event, termed the 'Najd event' between $\sim$680 and 630 million years ago, is recorded across Arabia. Pan-African terrane boundaries form steeply dipping zones that appear to have controlled subsequent basin evolution, particularly the South Oman and Ghaba salt basins.

The stratigraphy of the Precambrian and early Paleozoic rocks in Oman is shown in Fig. 3.2. Most of the stratigraphic record comes from oil wells along the

© Springer Nature Switzerland AG 2019
M. Searle, *Geology of the Oman Mountains, Eastern Arabia*,
GeoGuide, https://doi.org/10.1007/978-3-030-18453-7_3

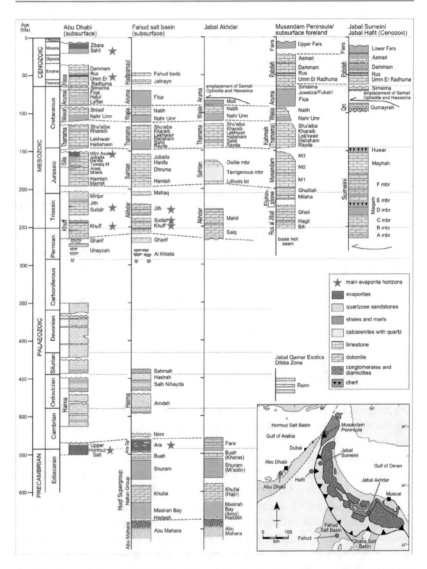

Fig. 3.1 Stratigraphy of UAE, Musandam peninsula, Northern Oman and the Fahud salt basin, after Cooper et al. (2018)

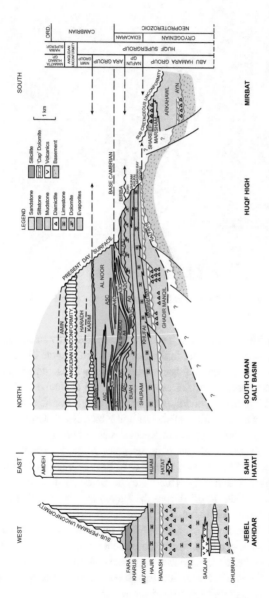

Fig. 3.2 Stratigraphic chart showing the Precambrian and early Paleozoic rocks of the northern Oman Mountains in Jebel Akhdar and Saih Hatat, South Oman and Dhofar, courtesy of Bruce Levell

Fig. 3.3 Neoproterozoic cap dolomite to the Precambrian glacial rocks, Wadi Hajir (photo courtesy of Bruce Levell)

flanks of the Huqf high. Proterozoic basement is exposed only in Jebel Ja'alan in the far northeast of the country and in the Darbat area of Dhofar. Here, a network of basaltic dykes has intruded felsic and mafic gneisses, amphibolites, and meta-diorites. These basement rocks extend east to the Hallaniyat (Kuria Muria) islands. A period of transtension followed during the Infracambrian (~560–540 million years ago) with formation of the Rub al-Khali, South Oman, Fahud and Ghaba salt basins (Ara Group in south; Hormuz salt in north).

Neoproterozoic-Cambrian glacial deposits (Abu Muhara Group) are seen in outcrop in the Huqf and South Oman as well as in Precambrian rocks in the core of Jebel Akhdar (Figs. 3.3 and 3.4). These are the outcrops that suggest that ice covered most if not all the Earth during the Proterozoic 'snowball earth' event. Many researchers have concluded that there were two separate main glacial intervals at ca. 720–700 Ma and ca. 645–635 Ma. Oman was in equatorial latitudes at this time and if thick glacial deposits are present here, then it is likely that ice covered the whole globe, either in one event or several different periods separated by short inter-glacial periods. Neoproterozoic Snowball Earth glacial deposits have been found in numerous localities around the world.

Fig. 3.4 Neoproterozoic Mistal Formation glacial sediments (Snowball Earth) in core of Jebel Akhdar, Wadi Mistal. In the background the prominent mid-Permian unconformity can be seen overlying the Precambrian and Paleozoic rocks (photo courtesy of Bruce Levell)

3.2 Proterozoic Sediments and Salt Basins

Deepening of the four major salt basins throughout the Late Proterozoic–Early Cambrian resulted in deposition of a great thickness of evaporites (Ara Group) accumulating in each basin. The South Oman and Ghaba salt basins are aligned NNE–SSW to the west of the Haushi–Huqf high running from Dhofar in the south up to the Oman mountain front in the north. The Fahud salt basin to the west separated from the Ghaba basin by the Makarem high also extends northward to the mountain front. Recent discoveries of Precambrian Ara salt extruded upwards along faults in tectonic windows beneath the obducted Semail ophiolite at Jebel Qumayrah and the Hawasina Window (Cooper et al. 2012, 2013, 2018; Ali et al. 2014) require the Fahud salt basin to have extended northwards at least as far as the Oman Mountain front (Fig. 3.5).

Six Precambrian salt domes pierce the surface in central Oman: Qarn Sahmah, Qarn Nihayda, Qarat Kibrit, Qarn Alam, Jebel Majayiz and Qarat al Milh (Peters et al. 2003). Five cycles of evaporites—carbonates make up the Ara Group, each

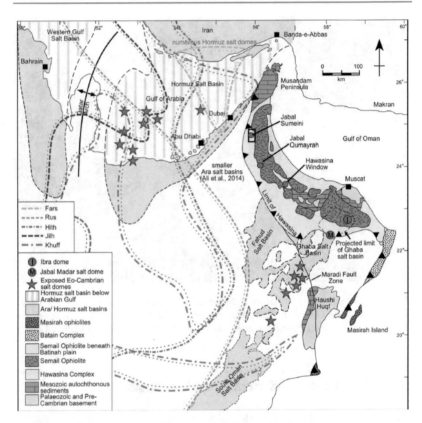

Fig. 3.5 Map of Eastern Arabia showing extent of Precambrian Hormuz-Ara salt basins and the South Oman, Fahud and Ghaba salt basins, after Cooper et al. (2018)

salt layer capped by a limestone layer between 10 and 150 m thick. The limestones, sometimes containing stromatolites, or algal mats, thought to be some of the oldest known forms of life, form great stringers or exotic blocks brought up by salt diapirism. Single-cell stromatolites are found throughout Late Proterozoic rocks in Oman and layers or mounds of these fossil bacterial algae are exposed in the Huqf region, above many of the Central Oman salt domes and in Jebel Akhdar. Sponges are the oldest known multi-cell organisms and the oldest fossils anywhere in the World were found in sedimentary rocks 635 million years old from the Huqf region of Oman (Al-Kindy 2018).

Some salt diapirs extend continuously from 8 to 10 km depth, from the main Precambrian salt horizon to the surface; others are detached salt pillows intruding to the surface. In the UAE a cluster of Hormuz salt domes reach the surface offshore Abu Dhabi including the islands of Sir Bani Yas, Delma, Arzana, Qarnain, Zirku and Sir Bu N'Air and one, Jebel Dhanna crops out along the coast west of Abu Dhabi (Thomas et al. 2015). Jebel Dhanna and Sir Bani Yas salt domes show Hormuz salt breccias with clasts of sedimentary rocks and volcanic rocks that have U–Pb zircon ages of 560–545 Ma (Thomas et al. 2015). These are a part of the massive Hormuz salt basin with several hundred salt domes intruding the Zagros fold-thrust belt in SW Iran and around the Qatar peninsula.

An Early Cambrian transpressional event, termed the 'Angudan event' ($\sim$540–520 million years ago) coincided with the ending of the Pan-African orogeny, formation of major strike-slip faults (e.g. Maradi fault in Central Oman), and important clastic sedimentary input into the Rub al-Khali basin. Reactivation of older terrane boundaries resulted in normal faulting along the flanks of the South Oman and Ghaba salt basins. The oldest oil and gas reservoirs in the World occur in the Precambrian (Neoproterozoic) Shuram, Buah and Ara Formations in south Oman. Further oil reservoirs in south Oman occur in the Lower Ordovician Haima Group and the Upper Carboniferous–Lower Permian Al Khlata Formation (Levell et al. 1988).

3.3 Paleozoic Sequence

Moderately stable platform sedimentation occurred from Late Cambrian to Late Carboniferous. Early Cambrian continental clastic sediments form a major oil and gas reservoir in south Oman (Nimr Formation). Paleozoic sequences are dominated by marine clastic sedimentation derived from the interior of Arabia from the west and south; carbonates only became dominant after the Middle Permian rifting (Konert et al. 2001). Major hydrocarbon reserves have been discovered in the Cambro-Ordovician section in South Oman, notably in the Ordovician Barik sandstone and the overlying Gharif horizon.

Massive Early Cambrian sandstones were followed by thin Middle Cambrian carbonates and then prograding delta sequences. The end of the Ordovician is marked by a regional unconformity formed during a large drop in sea-level, the result of another glaciation event. When the ice melted during subsequent global warming, a massive flooding event is recorded in Silurian rocks, mainly shale sequences that form an important oil source horizon (Qusaiba Formation in Saudi Arabia; Sahmah Formation in South Oman). The Late Silurian to Late Carboniferous is poorly represented in the Oman subsurface record mainly because of orogenic deformation and uplift during the assembly of Gondwana.

Following the Cambrian 'explosion' plants and animals evolved rapidly. Brachiopods, bivalves, echinoderms, crinoids and corals in shallow marine sediments and graptolites, primitive fish and nautiloids occur in deeper marine sediments. Trilobites are found commonly together with their trace fossils, *Cruziana* trails. In the oceans, microscopic, single celled foraminifera with calcite shells and radiolaria with silica shells bloomed; these micro-fossils, together with conodonts and ostracods are commonly used as stratigraphic markers throughout the Phanerozoic rock record. Oman also contains the earliest known land plants, 475 million years old, found in Ordovician rocks from central Oman (Al-Kindy 2018).

During the Late Carboniferous another orogenic cycle began during a major phase of orogenic compression at ∼315–295 million years ago. This event resulted in deformation with folding and thrusting of Paleozoic rocks and strong cleavage development as seen in Jebel Akhdar, weak metamorphism as seen in Saih Hatat, and uplift and erosion of north and east Oman. A major phase of halokinesis (salt diaprism) also occurred in North Oman with several major hydrocarbon traps above basement faults. The Late Carboniferous–Early Permian was a third important period of intense glaciation with deposition of the Al Khlata Formation glacial tillites, diamictites and conglomerates across northern and central Oman. Glacial striations on bedrock pavements, tillites and diamictites occur along the western flank of the Huqf anticline. The end of the glaciations resulted in a global rise in sea-level and drowning of the Arabian continental margin. In south Oman post-glacial clastic deposits occur in the Gharif Formation. This drowning event is seen in the Early Middle Permian siliciclastic and carbonate deposits in the Haushi–Huqf area (Saiwan Formation), and in the earliest units of the Hawasina complex in the northern Oman mountains as well as the Batain mélange along the east coast of Oman.

The boundary between the Permian and Triassic is the largest known period of mass extinctions on Earth when approximately 95% of marine species and 70% of terrestrial vertebrates became extinct, 252 million years ago. The cause is thought be a combination of meteorite impacts, and a period of widespread volcanism (Siberian Traps, and Emieshan Traps in China).

3.4 Middle Permian–Cretaceous Shelf Sequence

The break-up of the Gondwana supercontinent resulted in the rifting of the Arabian plate and establishment of a large-scale passive shelf margin in the Middle Permian ($\sim$270–255 million years ago). The earliest rifting deposits along the shelf margin are dated as Early Permian in the Jebel Qamar 'exotic' in the Dibba zone (UAE) and mid-Permian (Saiq Formation) around the Jebel Akhdar and Saih Hatat shelf carbonate culminations (Glennie et al. 1973, 1974). The Jebel Qamar 'exotic' in the Dibba zone is the only exotic in Oman to show a pre-Permian basement (Ordovician Rann Formation, and Lower Carboniferous Ayim Formation) indicating that it was a rifted block of the Arabian margin (Hudson et al. 1954; Pillevuit et al. 1997). Whereas in Jebel Akhdar, the Permian is inner shelf facies, around Saih Hatat the facies it is more outer-shelf and continental slope. Rifting of Gondwana formed a continental margin along northeast Arabia with a shelf—basin transition as the Tethyan ocean widened. Rocks formed outboard of the continental margin were deposited in the Hawasina basin, and are preserved in a series of thrust sheets on top of the Mesozoic carbonates (Fig. 3.6).

The Middle and Late Permian '*Fusulinid* Sea' transgressed over the Arabian plate, with the exception of the sub-aerial Haushi–Huqf 'high' and its northern extension, the Jebel Ja'alan massif. In Jebel Akhdar (Fig. 3.7) the Middle and Late Permian shelf carbonates comprise a 400 m thick succession (Saiq Formation) of shallow marine carbonates, and in interior Oman in the south (Khuff Formation). A notably thicker section (750 m) is exposed in the Musandam peninsula in the north in the Bih and Hagil Formations. In Saih Hatat a similar thickness of Saiq Formation limestones overlies a thin volcanic unit (Sq1) of tuffs, breccias and dacitic volcanics. The Saiq Formation overlies all earlier Palaeozoic units with a major angular unconformity seen all around the Jebel Akhdar and Saih Hatat windows. Stable shelf carbonate deposition lasted more than 160 million years is recorded by approximately 4–5 km thickness of limestones beautifully exposed in wadis draining off the Jebel al Akhdar (Fig. 3.8).

This widespread continental rifting event was accompanied by alkali basaltic intrusions in the Tethyan oceanic sequences (Hamrat Duru Group and distal Hawasina thrust sheets) as well as along the Batain coast, the eastern continental margin of Arabia. In the Haybi complex Upper Permian oceanic guyots or seamounts associated with an alkali basalt substrate are found in the highest thrust sheets beneath the Semail ophiolite (Haybi complex; Fig. 3.9). Some exotics, such as Jebel Misfah (Fig. 3.10) are more akin to rifted segments of the Arabian platform or shelf carbonates, rather than isolated ocean islands. Continued rifting and

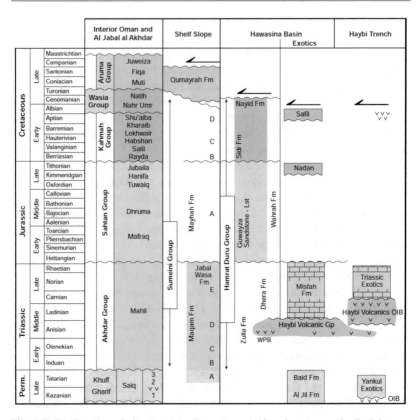

Fig. 3.6 Stratigraphy of the Permian–Cretaceous shelf carbonates, and allochthonous Hawasina and Haybi complexes, UAE-Oman Mountains, after Searle (2007)

passive margin carbonate deposition continued throughout the Upper Permian (Gharif and Khuff Formations) and Triassic (Mahil Formation in Oman, Ghail Formation in Musandam). Important oil reservoirs are present in the Permian Gharif sandstones in the Ghaba structure in south Oman. The Khuff evaporites are the World's largest reservoir of gas covering Saudi Arabia, Qatar, Abu Dhabi, and Oman. Four reservoir horizons each comprise cycles that begin with transgressive oolitic dolomites and mudstones and end with evaporate—anhydrite caps. The Triassic sequence was characterised by quiescent shallow marine or inter-tidal conditions, with the characteristic thick-shelled *Megalodon* sp. bivalves (Fig. 3.11). In Musandam the Upper Triassic–Lower Lias Ghalilah Formation is

Fig. 3.7 Permian and Mesozoic shelf carbonate sequence **a** north face of Jebel Shams, and **b** Wadi Nakhr Grand Canyon

Fig. 3.8 Jurassic and Cretaceous shelf carbonates, Wadi Nakhr, Jebel Akhdar

Fig. 3.9 Jebel Misht 1000 meter high cliffs of Upper Triassic exotic reefal limestones on top of a thin alkali basalt substrate

Fig. 3.10 Jebel Misfah Upper Triassic exotic limestones thrust over Hawasina cherts and Cretaceous shelf carbonates, Jebel Akhdar

Fig. 3.11 Triassic *Megalodon* sp. bivalves in Oman Exotic limestones

some 250 m thick and composed of distinctive orange-brown sandstones and shales containing thin-bedded brachiopod and gastropod-rich faunas.

From Middle Permian to Cenomanian time, 160 million years of extremely stable shelf carbonate deposition occurred across the northeastern continental margin of Arabia. During this time rifting resulted in a wide Tethys ocean separating the southern Gondwana continents to the south (Africa, Arabia, India) from the European and Asian landmass to the north. Marine transgressions across the Arabian shelf continued throughout the Jurassic (Sahtan Group) and Early Cretaceous (Musandam Group in the north; Kahmah Group in Jebel Akhdar and Saih Hatat) with near continuous carbonate deposition over a wide area of eastern Arabia. These carbonate reservoirs with evaporite seals form the richest oil habitat in the World across Saudi Arabia and UAE, with the enormous reserves in the Arab A to D reservoirs (Alsharhan and Kendall 1986). These limestones are rich in fossils with abundant foraminifera, echinoderms, belemnites, bivalves (*Lithiotis* sp.) and gastropods (Pratt and Smewing 1990; van Buchem et al., 2002). Two deepening events recorded in the mid-Jurassic and end-Jurassic (Rayda Formation cherts; 142–138 million years) interrupted an otherwise continuous shallow marine carbonate deposition.

Following the deepening event recorded at the Jurassic-Cretaceous boundary, a return to shallow marine conditions is seen in the overlying Salil and Habshan Formations with progradation of the shelf (Fig. 3.12). The overlying Lekhwair–Kharaib–Shuaiba Formations (Hauterivian to Aptian; 131–115 million years) record several shoaling upward cycles. The top Shuaiba Formation is marked by a regional unconformity or disconformity at around 113 million years ago, a result of a world-wide rise in sea-level (Haq et al. 1987). A return to stable shelf carbonate deposition is marked by onlapping shallow water conditions with abundant benthic foraminfera particularly *Orbitolina* sp. of the Aptian Nahr Umr and Cenomanian Natih Formations (Scott 1990), or Mauddud Formation in Musandam. The Albian-Aptian to Cenomanian (Cretaceous) shelf carbonates of the Natih and Mauddud Formations are major hydrocarbon reservoirs in northern Oman and UAE. The limestones are rich in micro-fossils especially foraminifera, as well as gastropods, bivalves, belemnites, multi-celled corals and single cell rudists (Al-Kindy 2018).

An abrupt end to stable shelf carbonates occurred after deposition of the Wasia Group limestones at ∼92–91 million years ago, an event termed the Wasia-Aruma hiatus (or 'break'). During the Late Cenomanian to Campanian ∼95–81 Ma the shelf margin collapsed, and a foreland basin, the Aruma foredeep developed along the margin to accommodate the incoming allochtonous thrusts sheets of the Tethyan oceanic sequence Hawasina, Haybi and Semail ophiolite thrust sheets

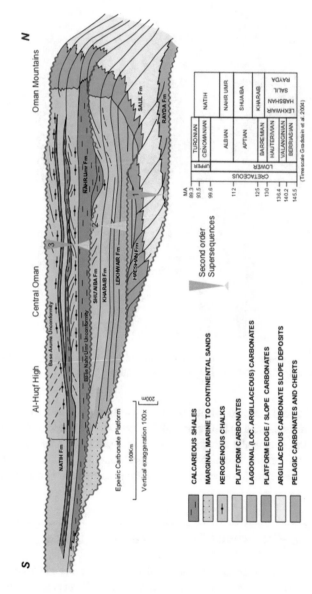

Fig. 3.12 Stratigraphy and facies correlation of the Cretaceous shelf carbonate sequence, after Droste (2005). Courtesy Henk Droste

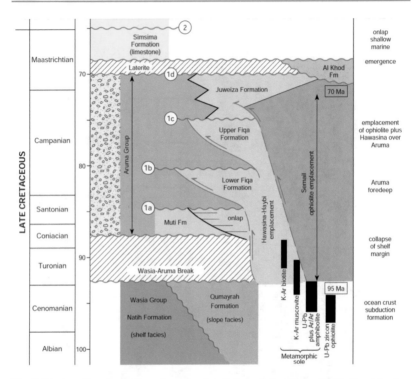

Fig. 3.13 Late Cretaceous time chart showing timing relationships of ophiolite obduction events, and Aruma foreland basin with major unconformities

(Fig. 3.13). The end-Cenomanian marks a major tectonic break with initial emplacement of the Semail ophiolite and all Tethyan oceanic rocks between the ophiolite and the continental margin, thrust from NE to SW into the rapidly deepening Aruma flexural basin. Numerous unconformities within the Aruma basin (Juweiza Formation) have been recognised from seismic and well data in the foreland basin.

3.5 Permian–Cretaceous Tethyan Basin Sequence

The earliest sedimentary rocks in the continental slope facies Sumeini Group in the Jebel Sumeini area are Middle Permian–Triassic Maqam Formation (Watts and Garrison 1986; Watts 1990), equivalent to the Khuff–Saiq Formation on the autochthonous shelf. The oldest radiolarian cherts in the Hawasina complex are also Middle Permian (De Wever et al. 1988; Cooper 1990; Baud et al. 2010), suggesting that the shelf-slope-basin distinction along the Oman margin was already in place by this time. The Tethyan oceanic sediments now preserved in the allochthonous thrust sheets structurally above the shelf carbonates and beneath the Semail ophiolite are time-equivalent units (Middle Permian to Cenomanian) as the authochthonous shelf carbonates (Glennie et al. 1973, 1974; Searle et al. 1983). From base to top these include the proximal continental shelf margin carbonates (Sumeini Group), continental slope (Hamrat Duru Group), distal (Hawasina complex) and trench (Haybi complex) rocks. These stratigraphic units are preserved within each structural thrust sheet carried on the basal Sumeini thrust, Hawasina thrust and Haybi thrust respectively (Fig. 3.14). These thrust sheets can be restored to reveal a shelf—slope—basin facies extending hundreds of kilometres offshore from the continental margin to the distal parts of the Tethys ocean (Glennie et al. 1974; Searle et al. 1983; Searle 2014).

The most proximal thrust sheet is the Sumeini thrust sheet composed of Middle Permian–Triassic Maqam Formation and Jurassic–Lower Cretaceous Mayhah Formation (Watts 1990). These are slope facies carbonates deposited on a carbonate ramp outboard of the autochthonous shelf, with conglomerates and occasional chert horizons correlating with the Jurassic-Cretaceous drowning event on the platform. During the Cenomanian–Coniacian a major deepening event is recorded by siliceous radiolarian mudstones and breccias. This is interpreted as the collapse of the shelf margin with the beginning of thrusting of the Tethyan outboard thrust sheets of the Hawasina and Haybi complexes.

The Hawasina complex rocks are base of slope and proximal basin deposits that show deeper water facies with increasing distance away from the Arabian continental margin. The Hamrat Duru Group includes the Permian–Triassic Al Jil Formation, the Jurassic Matbat Formation sandstone and Guwayza Formation limestone, the Sidr Formation cherts recording the Jurassic-Cretaceous drowning event further inboard, and the Cretaceous Nayid Formation (Cooper 1990; Bechennec et al. 1990). The Al Aridh Group and the Dibba and Shamal Formation cherts represent the distal Hawasina units. Whereas sedimentation on the continental shelf ceased abruptly at the Cenomanian–Turonian boundary sedimentation in the more distal units persisted until later.

The most distal units of the Tethyan thrust sheets beneath the ophiolite is the Haybi complex, a distinct thrust sheet comprised of Late Permian and Late Triassic exotic limestones interpreted as seamounts with an alkali basalt substrate, deep water cherts and mélanges formed in the trench immediately beneath the Semail ophiolite (Searle and Malpas 1980). These mélanges include olistostromes with

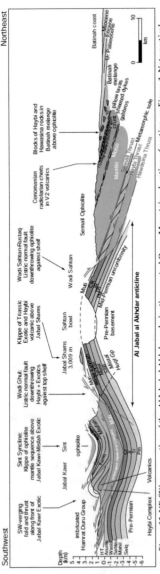

Figure 13: NE-SW cross-section of the Al Jabal al Akhdar culmination, central Oman Mountains; location of section on Figure 1. Note vertical exaggeration. Please see poster for true scale sections.

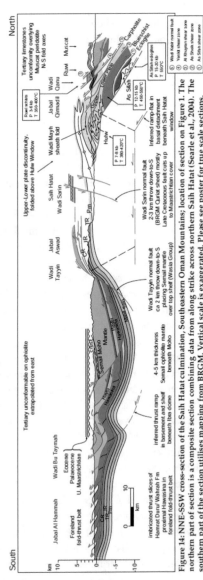

Figure 14: NNE-SSW cross-section of the Saih Hatat culmination, Southeastern Oman Mountains; location of section on Figure 1. The northern part of section is a composite section combining data from along strike across northern Saih Hatat (Searle et al., 2004). The southern part of the section utilises mapping from BRGM. Vertical scale is exaggerated. Please see poster for true scale sections.

Fig. 3.14 Geological cross-sections across Jebel Akhdar and Saih Hatat, after Searle (2007)

shaley matrix and serpentinite mélanges with serpentine matrix enclosing blocks of metamorphic sole rocks and all distal sedimentary lithologies.

Thrusting of these Tethyan oceanic rocks started 96–95 million years ago when the ophiolite crustal sequence formed above a NE-dipping subduction zone. Basalts were subducted to form granulite and amphibolite facies rocks that crystallised at depth and were accreted along the base of the ophiolite (metamorphic sole) at exactly the same time as when the ophiolite formed (96.0–95.5 million years ago; Rioux et al. 2016). Extrusion of the metamorphic sole from the subduction zone started the whole process of thrusting distal units over more proximal units as the Tethyan ocean closed and the shelf margin subsided to accommodate the incoming thrust sheets. Thrusting occurred in a piggy-back sequence from NE to SW with the Semail ophiolite thrust over the Haybi thrust sheet (Searle 1985). Both these units were then emplaced over the Hawasina thrust sheets and all of these were finally thrust over the shelf margin for at least 150 km across the drowned shelf carbonates of Jebel Akhdar and Saih Hatat.

3.6 Permian–Triassic Volcanism

Permian volcanic rocks occur within the Saiq Formation shelf carbonates (Sq1) in Saih Hatat, along the base of many of the Hawasina thrust sheets, and in the distal Haybi Complex (Searle et al. 1980; Maury et al. 2003). The lavas and flows within the shelf sequence and in the proximal Hawasina complex are all high-titanium alkaline lavas related to the rifting of the Arabian margin. True MORB volcanics are not present and it is unlikely that true MORB oceanic crust was formed until at least Late Triassic. Maury et al. (2003) related the Permian volcanism in Oman to a mantle plume, but the widespread distribution, extending over a distance of 350 km from the Dibba zone in UAE to the eastern Oman Mountains, and the paleogeographic distribution from the shelf margin outboard to Haybi complex precludes a mantle plume origin. It seems far more likely that the Permian volcanism was related to continental rifting and break-up, with more localized within-plate, off-axis volcanism related to ocean-island guyots formed off the passive continental margin during the late Permian and late Triassic (Glennie et al. 1974; Searle et al. 1980).

In the most distal Haybi complex thrust sheets nephelinites, ankaramites, alkali basalts and trachytes represent extremely alkaline lavas, with rare, but widespread intrusions of highly alkali peridotites (jacupirangites, biotite wehrlites) and kaersutite gabbros beneath exotic limestones (Searle 1984). These rocks are characteristic of present-day ocean islands located offshore passive continental margins, such as the Comores, Tristan da Cunha, Ascension Island, the Canary Isles etc.

A similar tectonic position is thought to be comparable for the origin of the Oman Haybi volcanic rocks. The Haybi volcanics formed the protolith of the sub-ophiolite metamorphic sole amphibolites (Searle and Malpas 1980, 1982). Thus, the footwall of the ophiolite was old, cold Triassic and Jurassic ocean floor (Searle and Cox 1999, 2002), not Cretaceous ocean floor equivalent to the Semail ophiolite (Boudier et al. 1988).

3.7 Oman Exotics: Baid and Misfah Platforms

Two types of 'Oman Exotic' limestones occur in the allochthonous thrust stack structurally above the shelf carbonates. The first group includes large-scale, well-bedded carbonate banks frequently with an alkali basalt basement. Examples are Upper Triassic exotics of Jebel Misfah, Jebel Kawr (Fig. 3.15) and Jebel Misht (Fig. 3.16) around the SW flank of Jebel Akhdar, and the Upper Permian Baid exotic, south of Saih Hatat. The second group includes numerous smaller exotics, frequently intimately associated with within-plate, highly alkaline volcanic rocks, often occurring in mélanges and imbricate slices immediately beneath the Semail ophiolite (Searle and Graham 1982; Pillevuit et al. 1997).

Fig. 3.15 Jebel Kawr Exotic view southwest from the summit of Jebel Shams

Fig. 3.16 Jebel Misht Exotic showing 1000 m of Upper Triassic reefal limestone overlying thin Triassic alkali basalts, thrust over red cherts of the distal Hawasina complex

The Baid platform carbonates comprise Late Permian platform limestones, Triassic (*Hallstatt* facies, red ammonite-rich limestones) overlain by thin cherty limestones and radiolarites of Early Jurassic age (Blendinger et al. 1990; Pillevuit et al. 1997). This platform sequence probably rifted away from the Oman continental margin as it occurs northeast, outboard of the true shelf and the proximal Hawasina rocks in the paleogeographic reconstruction (Bernoulli and Weissart 1987; Bernoulli et al. 1990). These exotics have been thrust to the SW above Hamrat Duru (Hawasina complex) thrust sheets that are structurally above the Cretaceous shelf and its cover of Aruma Group, and immediately beneath the Semail ophiolite.

In Jebels Misfah, Kawr and Misht, structurally above the Jebel Akhdar shelf sequence, 700–800 m thick Late Triassic platform limestones (with Late Dogger *Ammonico Rosso* facies) up to 1000 m thick, were deposited in 10–15 million years. These limestones are similar to the Triassic shelf (Mahil Formation), but overlie an alkaline basaltic substrate indicative of off-axis, within-plate ocean-island volcanic rocks (Searle et al. 1980). On the Jebel Misfah exotic, foundering of the Triassic seamount occurred during the Late Oxfordian to Tithonian (Jurassic) with deposition of thin deep-water pelagic limestones and cherts of the Nadan Formation at the top (Pillevuit et al. 1997). During the Late Triassic thermal subsidence following sea-floor spreading resulted in subsidence and drowning of the Triassic exotic carbonate platform.

These exotic limestones have been thrust over proximal Hamrat Duru Group and Al Aridh Formation (Hawasina complex) rocks and occur within the Haybi complex thrust sheet, immediately beneath the Semail ophiolite, indicating their palinspastic restoration position well outboard of the Arabian continental shelf and shelf margin.

3.8 Late Triassic Ocean-Island 'Exotic' Limestones

Numerous Late Triassic exotic limestones occur throughout the Haybi Complex thrust sheet immediately beneath the ophiolite (Glennie et al. 1973, 1974; Searle and Graham 1982). These rocks were later termed part of the Umar Group (BRGM mapping; Bechennec et al. 1990; Rabu et al. 1993); however, they span Late Permian to Middle Cretaceous, are time-equivalent rocks to the Hawasina and Sumeini thrust sheets and the autochthonous shelf, and are bounded by thrust faults, so the structural term 'Haybi Complex' is more appropriate. Many Triassic exotics show large spectacular *Megalodon sp.* bivalves within coral limestones, with flanking mega-breccias representing ocean-island slope deposits. Blocks of shallow-water Triassic exotic limestone are sometimes encased in deep-water radiolarian cherts indicating base of slope facies.

The most impressive of all Oman Exotics is Jebel Misht, formed as an isolated oceanic guyot or island out in the Tethys ocean, and now forming the thrust sheet immediately beneath the Semail ophiolite. The presence of alkali basalts, and some highly alkaline peridotite–gabbro intrusions into the Late Triassic exotic basements suggest that their origin was on a series of isolated within-plate, off-axis oceanic volcanic island seamounts. These Triassic oceanic floor rocks of the Haybi complex restore to a paleogeographic position immediately SW of the location where the subduction zone would later initiate beneath the Semail ophiolite.

3.9 Aruma Foreland Basin

Foreland basins develop due to flexure as a result of loading of an adjacent mass, usually a mountain belt. A large-scale gravity 'low' over the Oman Mountains foreland has been interpreted in terms of elastic flexure of the Arabian foreland beneath the allochthonous thrust sheets exposed in the mountains (Ravaut et al. 1997). The Late Cretaceous Aruma foredeep basin includes up to 3.5 km thickness of Santonian–Campanian Fiqa Formation with up to 1.5 km thickness of Late Campanian Juweiza Formation, thickening to the NE. Major changes in platform margin sedimentation occurred following the Cenomanian when the whole continental margin collapsed as recorded by large-scale megabreccias, intra-formational unconformities and transgression to deep water clastic sediments of the Qumayrah Formation along the margin and the time equivalent Aruma Group (Robertson 1987; Watts 1990). Collapse of the shelf margin and beginning of the foreland basin deposition following the 'Wasia-Aruma break' (Turonian) was accompanied by uplift along the peripheral bulge. Up to 1100 m of uplift and erosion of the Natih Formation on the Lekhwair High may have been caused by flexural bending of the foreland (Warburton et al. 1990; Loosveld et al. 1996; Filbrandt et al. 2006) along the Suneinah Trough (Patton and O'Connor 1988; Boote et al. 1990).

The Aruma (Suneinah) basin shows a rapid increase in thickness towards the NE with Coniacian-Santonian Muti Formation conglomerates passing up to deeper water shales and siltstones of the Campanian Fiqa Formation. The Late Campanian Juweiza Formation shows the first debris eroded from the Semail ophiolite (Glennie et al. 1974). Seismic sections across the foreland show the Hawasina thrust tips extending up into the Fiqa Formation (Warburton et al. 1990; Boote et al. 1990). The entire thrust process was completed by 70 Ma (Lower Maastrichtian) when the mountain belt just breached sea-level. Thin laterite soils overlie the ophiolite erosional surface in the southeast Oman Mountains.

Late stage folding of the giant Jebel Akhdar and Saih Hatat anticlines started during the final stages of ophiolite obduction (Fig. 3.14). Most of the uplift of Jebel Akhdar and Saih Hatat is Late Cretaceous in age but up to 2000 m of uplift was further enhanced by mid- to Late Cenozoic compression. Significant tectonism and uplift of the foreland region northeast of the Maradi fault is also known from well cores and seismic sections.

3.10   Maastrichtian–Cenozoic Sedimentary Sequence

The Late Cretaceous period of ophiolite obduction and thrusting of Neo-Tethyan oceanic rocks from NE to SW onto the drowned passive margin of Arabia ended in early Maastrichtian time. A regional unconformity truncates all allochthonous units as well as the shelf carbonates throughout the Oman–UAE Mountains (Fig. 3.17).

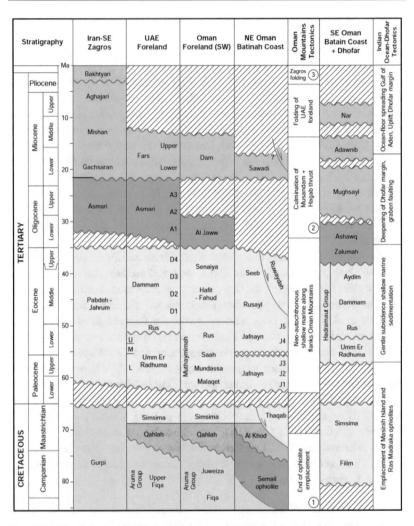

Fig. 3.17 Time chart showing Late Cretaceous–Cenozoic stratigraphy spanning Zagros Mountains, Iran, UAE, Oman Mountains and the Batain coast

Fig. 3.18 Death assemblage of rudists and gastropods in Maastrichtian Simsima Formation limestones, Jebel Sumeini

Above the unconformity conglomerates and shales of the Qahlah and Al Khod Formations are overlain by shallow water rudist-bearing limestones indicative of a return to stable shelf conditions (Simsima Formation). The abundance of branching corals, rudists, bivalves and gastropods in the Simsima Formation across all of north Oman is testament to the fact that tectonism had ceased by Maastrichtian time, 70 million years ago. Many Maastrichtian outcrops throughout the Oman Mountains show a spectacular 'death assemblage' of rudists, gastropods and other species as Mesozoic era came to an end (Fig. 3.18).

Life exploded soon after the end of the ophiolite obduction event in the Oman–UAE Mountains. Shallow marine limestones in the Maastrichtian are packed with rudists, branching corals, gastropods, bivalves and foraminifera (Figs. 3.19 and 3.20). Fossils of four different species of dinosaur (therapods, ornithopods, sauropods), turtles and numerous trees have been found in the Al Khod Formation

Fig. 3.19 Cenozoic gastropod fossils

(Al-Kindy 2018). Another major, global mass extinction occurred at the Creta-ceous–Cenozoic boundary at the end of the Maastrichtian 65 million years ago, when three-quarters of all plant and animal species on Earth became extinct, including the dinosaurs. Similar to the end-Permian extinction, the causes are thought be a giant meteorite impact (the Chicxulub crater in the Gulf of Mexico) and widespread volcanism, notably the Panjal Trap volcanism in India. The K-P boundary is marked by high levels of iridium, a metal that is rare on Earth, but abundant in asteroids.

North Oman became emergent again at the end of the Maastrichtian, due largely to a global eustatic fall in sea-level. Minor deformation and uplift of the mountains is indicated by the outcrop pattern of the two unconformities below and above the Simsima Formation. During the Late Paleocene stable shallow-water limestone deposition resumed and lasted for 25 million years until the Oligocene (Nolan et al. 1990; Mann et al. 1990). Life blossomed during this period both on land with the

Fig. 3.20 *Nummulites* foraminifera, Seeb Formation Eocene limestones, Al Khod, Batinah plain

dominance of mammals, and in the oceans. Paleocene–Eocene limestones throughout Oman and UAE are packed with foraminifera, as well as larger fossils, particularly corals, bivalves, echinoids, gastropods and fishes.

A second orogenic phase began after the Late Oligocene–Miocene, possibly extending up to Pliocene, with renewed crustal shortening and folding of the Paleogene limestones and underlying rocks. The style of folding in the Cenozoic is generally large-scale open folds or box folds with kink band sets. These folds may have steep, even overturned limbs but flat tops, characteristic of box folds with prominent kink-band sets, some of which may evolve into steep thrust faults. Shortening is minimal, but uplift of at least 2000 m is known from Jebel Abiad and Jebel Bani Jabir (Salma plateau), east of Saih Hatat (Fig. 3.21). The Abat basin south of Sur is a large embayment in the Cenozoic shelf margin and accumulated deeper water sediments as young as Upper Miocene. Large-scale folding of the

Fig. 3.21   Jebel Abiad, Fins-Ash Shab region showing Cenozoic limestones, eastern Oman. Note the uplifted marine terraces behind the beach

Paleocene–Miocene sediments of the Abat basin suggests that this stage of folding could be as young as Pliocene–Recent, similar in age to the Zagros folds of Iran. The cause of this compression in eastern Oman remain a mystery as no continental collision has occurred in this region of Eastern Arabia.

The magnitude and extent of Cenozoic folding increases to the north along the Oman foreland (Fig. 3.22). The Musandam peninsula shows large-scale re-folding and re-thrusting of the entire Pre-Permian basement up to Paleogene cover with minimum east to west translation of 14 km along the Hagab thrust (Searle et al. 1983; Searle 1988a, b; Ali et al. 2014). These west-vergent thrusts associated with the Hagab thrust are truncated by an Upper Miocene unconformity in seismic sections west of Musandam in the Arabian Gulf. These structures are interpreted as the beginning of the Zagros phase of continent–continent collision when the Musandam shelf first collided with the central Iran continental block (eastward extension of the Zagros suture zone).

Fig. 3.22 Paleocene-Eocene limestones around Bandar Jissah unconformably above mantle sequence peridotites of the Muscat–Muttrah ophiolite in distance

The pattern of fold axes along the northern mountains reflects a dome and basin fold interference pattern with a dominant NE–SW compressive stress resulting in the roughly east-west oriented folds of Jebel Akhdar and Saih Hatat and a subordinate WNW–ESE compressive stress resulting the NNE aligned fold such as Jebel Nakhl (Fig. 3.23). The Late Cretaceous NE–SW compression was concomitant with the ophiolite obduction event, but the cause of the WNW–ESE compression remains uncertain. It could have been the result of westward obduction of the Batain mélange and the Masirah ophiolite, although little effects of this process are known west of the Huqf–Jebel Ja'alan 'high'. It is unlikely to have been a sideways collision of the Indian plate during its rapid northward drift, as a significant extent of Indian Ocean crust lies west of the Owen Fracture zone between India and Arabia.

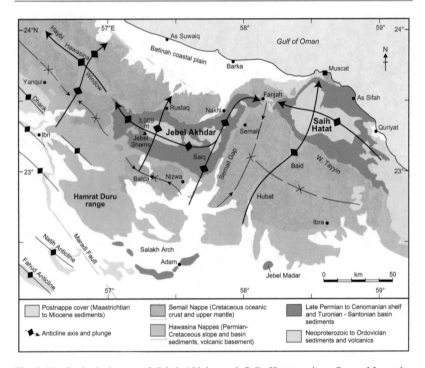

Fig. 3.23 Geological map of Jebel Akhdar and Saih Hatat region, Oman Mountains showing major fold axes and dome and basin fold interference pattern, after Searle (2007)

References

Alsharhan AS, Kendall CG (1986) Precambrian to Jurassic rocks of the Arabian Gulf, and adjacent areas: their facies, deposition setting and hydrocarbon habitat. Am Assoc Petrol Geol Bull 78:1075–1096

Al-Kindy MH (2018) Evolution of land and life in Oman: An 800 million year story. Springer, 220 p

Ali MY, Cooper DJW, Searle MP, Al-Lazki A (2014) Origin of gypsiferous intrusions in the Hawasina Window, Oman Mountains: implications from structural and gravity investigations. GeoArabia 19:17–48

Baud A, Bernecker M et al. (2010) The Permian-Triassic transition in the Oman mountains. IGCP Field Guide Book 2

Bechennec F, LeMetour, J, Rabu D, Bourdillon-de-Grissac C, DeWever P, Beurrier M, and Villey M (1990) The Hawasina Nappes: stratigraphy, palaeogeographyand structural evolution of a fragment of south Tethyan passive continental margin. In: Robertson AFH,

Searle MP, Ries AC (eds) The geology andtectonics of the Oman region. Geological Society, London, Special Publication no 49, pp 213–223

Bernoulli D, Weissert H (1987) The upper Hawasina nappes in the central Oman Mountains: straigrpahy, palinspastics and sequence of nappe emplacement. Geoldinamica Acta 1:47–58

Bernouli D, Weissart H, Blome CD (1990) Evolution of the Triassic Hawasina Basin, Central Oman Mountains. In: Robertson AFH, Searle MP, Ries AC (eds) The geology and tectonics of the Oman Region. Geological Society, London, Special Publication no. 49, pp 189–202

Blendinger W, Van Vliet A, Hughes Clarke MW (1990) Updoming, rifting and continental margin development during the late Palaeozoic in northern Oman. In: Robertson AFH, Searle MP, Ries AC (eds) The geology and tectonics of the Oman region. Geological Society, London, Special Publication no 49, pp 27–37

Boote D, Mou D, Waite RL (1990) Structural evolution of the Suneinah foreland, central Oman. In: Robertson AFH, Searle MP, Ries AC (eds) The geology and tectonics of the Oman Region. Geological Society, London, Special Publication no. 49, pp 397–418

Boudier F, Ceuleneer G, Nicolas A (1988) Shear zones, thrusts and related magmatism in the Oman ophiolite: initiation of thrusting on an oceanic ridge. Tectonophysics 151:275–296

Cooper DJW (1990) Sedimentary evolution and palaeogeographical reconstruction of the Mesozoic continental rise in Oman: evidence from the Hamrat DuruGroup. In: Robertson AFH, Searle MP, Ries AC (eds) The geology and tectonics of the Oman region. Geological Society, London, Special Publication no 49, pp 161–188

Cooper DJW, Searle MP, Ali MY (2012) Structural evolution of Jabal Qumayrah: a salt-intruded culmination in the northern Oman Mountains. GeoArabia 17:121–150

Cooper DJW, Ali MY, Searle MP, Al-Lazki AI (2013) Salt intrusions in Jabal Qumayrah, northern Oman Mountains: implications from structural and gravity investigations. GeoArabia 18:141–176

Cooper DJW, Ali MY, Searle MP (2018) Origin and implications of a thrust-bound gypsiferous unit along the western edge of Jabal Sumeini, northern Oman Mountains. J Asian Earth Sci 154:101–124

De Wever P, Bordillon-de-Grissac C, Bechennec F (1988) Permian age from radiolarites of the Hawasina nappes. Geology 16:912–914

Droste H (2005) Stratigraphic evolution and stratal geometries of the Cretaceous carbonate platform in Oman. Geological Society of Oman, Field Guide 014

Filbrandt JB, Nolan SC, Ries AC (1990) Late Cretaceous and early tertiary evolution of Jebel Ja'alan and adjacent areas, NE Oman. In: Robertson AFH, Searle MP, Ries AC (eds) The geology and tectonics of the Oman Region. Geological Society, London, Special Publication no. 49, pp 697–714

Filbrandt JB, Al-Dhahab S, Al-Habsy A, Harris K, Keating J, Al-Mahruqi S, Ozkaya I, Richard PD, Robertson T (2006) Kinematic interpretation and structural evolution of North Oman, Block 6, since the Late Cretaceous and implications for timing of hydrocarbon migration into Cretaceous reservoirs. GeoArabia 11(1):97–139

Glennie KW, Boeuf MG, Hughes-Clarke MHW, Moody-Stuart M, Pilaar WF, Reinhardt BM (1973) Late Cretaceous nappes in the Oman Mountains and their geologic evolution. Bull Am Assoc Pet Geol 57:5–27

Glennie KW, Boeuf MG, Hughes-Clarke MHW, Moody-Stuart M, Pilaar WF, Reinhardt BM (1974) Geology of the Oman Mountains. Verhandelingen Koninklijk Nederlands geologisch mijnbouwkundidg Genootschap 31:423p

Haq BU, Hardenbol J, Vail PR (1987) Chronology of fluctuating sea levels since the Triassic. Science 235:1156–1167

Hudson RGS, McGuigan A, Morton DM (1954) The structure of the Jebel Hagab area, Trucial Oman. Q J Geol Soc, London 110:121–152

Hughes-Clark M (1990) Oman's geological heritage. Petroleum Development Oman Ltd., 247 p

Konert G, Al-Hajri S, Al Naim A, Afifi AM, de Groot K, Droste HJ (2001) Paleozoic stratigraphy and Hyrdrocarbon habitat of the Arabian Plate. Ch. 24, Petroleum Provinces of the twenty-first century. Am Assoc Pet Geol Mem 74:483–515

Levell BK, Braakman JH, Rutten KW (1988) Oil-bearing sediments of Gondwana Glaciation in Oman. Am Assoc Petrol Geol Bull 72:775–796

Loosveld RJH, Bell A, Terken JJM (1996) The tectonic evolution of interior Oman. GeoArabia 1(1):28–51

Mann A, Hanna SS, Nolan SC (1990) The post-Campanian tectonic evolution of the Central Oman Mountains: tertiary extension of the Eastern Arabian margin. In: Robertson AFH, Searle MP, Ries AC (eds) The geology and tectonics of the Oman Region. Geological Society, London, Special Publication no. 49, pp 549–563

Maury R, Bechennec F, Cotton J, Caroff M, Cordey F, Marcoux J (2003) Middle Permian plume-related magmatism of the Hawasina Nappes and the Arabian platform: implications on the evolution of the Neotethyan margin in Oman. Tectonics 22:1073. https://doi.org/10.1029/2002TC001483

Nolan SC, Skelton PW, Clissold BP, Smewing JD (1990) Maastrichtian to early Tertiary stratigraphy and paleogeography of the central and Northern Oman Mountains. In: Robertson AFH, Searle MP, Ries AC (eds) The geology and tectonics of the Oman Region. Geological Society, London, Special Publication no. 49, pp 495–520

Patton T, O'Connor SJ (1988) Cretaceous flexural history of northern Oman Mountain foredeep, United Arab Emirates. Am Assoc Pet Geol 72:797–809

Peters JM, Filbrandt JB, Grotzinger JP, Newall MJ, Shuster MW, Al-Siyabi HA (2003) Surface-piercing salt domes of interior North Oman and their significance for the Ara carbonate "stringer" hydrocarbon play. GeoArabia 8:231–270

Pillevuit A, Marcoux J, Stampfli G, Baud A (1997) The Oman Exotics: a key to the understanding of the Neotethyan geodynamic evolution. Geodin Acta 10(5):209–238

Pratt BR, Smewing JD (1990) Jurassic and early Cretaceous platform margin configuration and evolution, central Oman Mountains. In: Robertson AFH, Searle MP, Ries AC (eds) The geology and tectonics of the Oman Region. Geological Society, London, Special Publication no. 49, pp 69–88

Rabu D, Nehlig P, Roger J, Béchennec F, Beurrier M, LeMetour J, Bourdillon-de-Grissac CH, Tegyey M, Chauvel J, Cavelier C, Al-Azri H, Juteau T, Janjou D, Lemiere B, Villey M, Wyns R (1993) Stratigraphy and structure of the Oman Mountains. Document BRGM, no 221, Orleans, France

Ravaut P, Bayer R, Hassani R, Rousset D, Al Yahya'ey A (1997) Structure and evolution of the northern Oman margin: gravity and seismic constraints over the Zagros–Makran–Oman collision zone. Tectonophysics 279, 253–280

Rioux M, Garber J, Bauer A, Bowring S, Searle MP, Keleman P, Hacker B (2016) Synchronous formation of the metamorphic sole and igneous crust of the Semail ophiolite: new constraints on the tectonic evolution during ophiolite formation from high-precision U–Pb zircon geochronology. Earth Planet Sci Lett 451:185–195

Robertson A (1987) The transition from passive margin to an Upper Cretaceous foreland basin related to ophiolite emplacement in the Oman Mountains. Geol Soc Am Bull 99:633–653

Robertson AHF, Searle MP (1990) The northern Oman Tethyan continental margin. In: Robertson AFH, Searle MP, Ries AC (eds) The geology and tectonics of the Oman Region. Geological Society, London, Special Publication no. 49, pp 3–25

Scott RW (1990) Chronostratigraphy of the cretaceous carbonate shelf, southeastern Arabia. In: Robertson AFH, Searle MP, Ries AC (eds) The geology and tectonics of the Oman Region. Geological Society, London, Special Publication no. 49, pp 89–108

Searle MP (1984) Alkaline peridotite, pyroxenite and gabbroic intrusions in the Oman Mountains. Can J Earth Sci 21:396–406

Searle MP (1985) Sequence of thrusting and origin of culminations in the northern and central Oman Mountains. J Struct Geol 7:129–143

Searle MP (1988a) Structure of the Musandam culmination (Sultanate of Oman and United Arab Emirates) and the Straits of Hormuz syntaxis. J Geol Soc, London 145:831–845

Searle MP (1988b) Thrust tectonics of the Dibba zone and the structural evolution of the Arabian continental margin along the Musandam Mountains (Oman and United Arab Emirates). J Geol Soc, London 145:43–53

Searle MP (2014) Preserving Oman's geological heritage: proposal for establishment of World Heritage Sites, National GeoParks and Sites of Special Scientific Interest (SSSI). In: Rollinson HR, Searle MP, Abbasi IA, Al-Lazki A, Al-Kindy MH (eds) Tectonic evolution of the Oman Mountains. Geological Society, London, Special Publication vol 392, pp 9–44

Searle MP (2007) Structural geometry, style and timing of defrmation in the Hawasina Window, Al Jabal al-Akhdar and Saih Hatat culminations, Oman Mountains. GeoArabia 12:93–124

Searle MP, Cox JS (1999) Tectonic setting, origin and obduction of the Oman ophiolite. Geol Soc Am Bull 111:104–122

Searle MP, Cox JS (2002) Subduction zone metamorphism during formation and emplacement of the Semail Ophiolite in the Oman Mountains. Geol Mag 139:241–255

Searle MP, Graham GM (1982) "Oman Exotics"—oceanic carbonate build-ups associated with the early stages of continental rifting. Geology 10:43–49

Searle MP, Malpas J (1980) The structure and metamorphism of rocks beneath the Semail Ophiolite of Oman and their significance in ophiolite obduction. Trans R Soc, Edinb Earth Sci 71:247–262

Searle MP, Malpas J (1982) Petrochemistry and origin of sub-ophiolite metamorphic and related rocks in the Oman Mountains. J Geol Soc Lond 139:235–248

Searle MP, James NP, Calon TJ, Smewing JD (1983) Sedimentological and structural evolution of the Arabian continental margin in the Musandam mountains and Dibba zone, United Arab Emirates. Geol Soc Am Bull 94:1381–1400

Searle MP, Lippard SJ, Smewing JD, Rex DC (1980) Volcanic rocks beneath the Semail Ophiolite in the northern Oman Mountains and their tectonic significance in the Mesozoic evolution of Tethys. J Geol Soc Lond 137:589–604

Sharland PR, Casey DM, Davies RB, Simmons MD, Sutcliffe OE (2004) Arabian plate sequence stratigraphy—revisions to SP2. GeoArabia 9(1):199–214

Thomas RJ, Ellison RA, Goodenough K, Roberts NMW, Allen PA (2015) Salt domes of the UAE and Oman: probing eastern Arabia. Precambr Res 256:1–16

Van Buchem FSP, Razin P, Homewood PW et al (2002) High-resolution sequence stratigraphy of the Natih Formation (Cenomanian/Turonian) in Northern Oman: distribution of source rocks and reservoir facies. GeoArabia 1:65–91

Warburton J, Burnhill TJ, Graham RH, Isaac KP (1990). In: Robertson AFH, Searle MP, Ries AC (eds) The geology and tectonics of the Oman Region. Geological Society, London, Special Publication no. 49, pp 419–428

Watts KF (1990) Mesozoic carbonate slope facies marking the Arabian platform margin in Oman: depositional history, morphology and palaeogeography. In: Robertson AFH, Searle MP, Ries AC (eds) The geology and tectonics of the Oman Region. Geological Society, London, Special Publication no. 49, pp 139–159

Watts KF, Garrison RE (1986). Sumeini group, Oman—evolution of a Mesozoic carbonate slope on a south Tethyan continental margin. Sediment Geol 48:107–168

Ophiolites and Regional Tectonics

4

Ophiolites are large-scale thrust sheets of oceanic crust and upper mantle rocks that have been emplaced (obducted) onto continental crust. The formal definition of an ophiolite complex dates back to the Penrose meeting in 1972 where it was decided that a typical 'ophiolite' comprises a mantle sequence ultramafic complex of peridotites at the base overlain by a crustal sequence made up of gabbros and a sheeted dyke complex with a mafic volcanic sequence, often pillow lavas, at the top. This Penrose ophiolite 'stratigraphy' became an analogue for the seismic velocity structure of the oceans, and was soon proven by an intense period of drilling during the Ocean Drilling Project (ODP). The later International Ocean Drilling Project (IODP) has spent millions of dollars drilling shallow wells in all the World's oceans, but these cores only ever penetrate the upper few kilometres of the crust down to the dyke-gabbro contact, and nowhere has a complete section through the oceanic crust been drilled. Ophiolite complexes therefore, are of immense importance because they reveal the petrological composition and structural architecture of the otherwise inaccessible oceanic crust and upper mantle.

Ophiolites, together with deep-sea sediments, high-pressure metamorphic rocks and mélanges mark ancient suture zones where continental plates have collided. The Semail Ophiolite exposed along the length of the Oman–UAE Mountains is well known as being the largest, most intact, best exposed and most intensely studied ophiolite complex anywhere in the World (Fig. 4.1). The ophiolite is exposed across the entire Oman–UAE Mountains, folded over the giant anticlines of Jebel Akhdar and Saih Hatat, and along the length of the mountains for over 600 km, at least from the Dibba zone in UAE in the northwest, to the Huqf–Jebel Ja'alan 'high' in northeastern Oman. Although the dark ophiolite hills are lower elevation than the massive shelf carbonates of the Musandam, Jebel Akhdar and Saih Hatat culminations, they are steep-sided and jagged with a unique 'moonscape' geomorphology. In Oman and UAE the Semail ophiolite forms the highest thrust sheet

© Springer Nature Switzerland AG 2019
M. Searle, *Geology of the Oman Mountains, Eastern Arabia*,
GeoGuide, https://doi.org/10.1007/978-3-030-18453-7_4

Fig. 4.1 View west across the Oman ophiolite to the Semail Gap and Jebel Nakhr. The Moho separates foreground gabbros of the lower crust from dark coloured harzburgites of the Mantle sequence

emplaced over a series of distal (Haybi complex) and proximal (Hawasina thrust sheets) sedimentary rocks that were formed in the Neo-Tethys Ocean. These allochthonous rocks were emplaced from NE to SW into a Late Cretaceous sub-siding foreland flexural basin (Aruma basin) that lies above the Permian–Mesozoic shelf carbonates and older rocks of the Arabian continental margin.

Oman has another, second ophiolite complex, the Masirah ophiolite located on Masirah Island off the coast of Al Wusta region in central Oman (Peters et al. 1997; Rollinson 2017). This is a different ophiolite from the Semail ophiolite in the northern mountains, having a Mid-Ocean Ridge Basalt (MORB) geochemistry and an older Jurassic age of formation. An upper sequence of Cretaceous lavas reveals an alkali Ocean Island basalt (OIB) geochemistry above the Jurassic ophiolite complex. Obduction processes are not well known because the base of the ophiolite is not exposed. Large oceanic strike-slip faults run along the southeast Arabian coast offshore Masirah and it is likely that some horizontal strike-slip motion combined with compression could have uplifted the Masirah ophiolite into its present position. The Masirah ophiolite is thought to have remained a part of the Indian ocean crust for about 80 million years prior to emplacement.

The major orogenic phase in north Oman and UAE occurred during the Late Cretaceous with emplacement of a series of thrust sheets of Tethyan oceanic rocks and the Semail ophiolite from NE to SW. This phase ended by the Maastrichtian approximately 70–68 million years ago, when shallow water carbonates were deposited on top of the thrust sheets. These Upper Maastrichtian limestones are packed with fossils, including large gastropods and rudist corals which died out at the World-wide, end-Cretaceous mass extinction event. A second phase of folding and crustal uplift occurred during the mid-Cenozoic and affected all units. In the north, the first effects of the continental collision (Arabian–Iran) are seen in the Musandam culmination, where thick-skinned thrusts affect basement rocks and all overlying units. Seismic evidence offshore west coast of Musandam and onshore in the UAE foreland show that these large-scale thrust faults are truncated unconformably by Middle Miocene rocks suggesting that the deformation event occurred during the period ∼25–15 million years ago.

4.1 Semail Ophiolite (Oman–UAE)

Since the earliest pioneering studies of Lees (1928), the first systematic mapping of the entire ophiolite (Reinhardt 1969; Glennie et al. 1973, 1974) and subsequent detailed geochemical studies (e.g. Pearce et al. 1981; Lippard et al. 1986) it has been widely recognised that the Semail Ophiolite is by far the most complete section though the oceanic crust and mantle preserved anywhere in the World (Fig. 4.2). Numerous later studies have mapped out the ophiolite and described the petrology, geochemistry and structure in great detail (e.g.: Coleman 1981; Hopson et al. 1981; Pearce et al. 1981; Lippard et al. 1986; Nicolas and Boudier 1995; Searle and Cox 1999; McLeod et al. 2013). U–Pb dating of zircons extracted from trondhjemites (plagiogranites) and gabbros in the lower crust parts of the ophiolite show that the Semail ophiolite formed during the Cenomanian at ∼96.12–95.50 million years (Rioux et al. 2013, 2016).

The Semail Ophiolite is unique for spectacular 3-dimensional outcrops and continuous exposure through the entire ophiolite sequence. The crustal section is between 6 and 8 km thick, overlying a Moho Transition Zone (MTZ) that can vary from 0 to 1 km thick, and an upper mantle section up to 15 km thick composed of ultramafic rocks (peridotites). Along the base of the ophiolite a Banded Ultramafic unit consists of sheared peridotites, including harzburgites, dunites and lherzolites, with up to 20% late hornblende (Ambrose et al. 2018; Ambrose and Searle 2019). These emplacement-related shear fabrics are related to early obduction and possibly to subduction initiation processes. The metamorphic sole was accreted to the

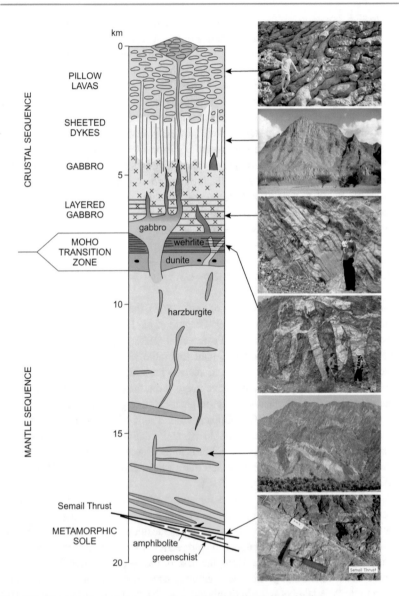

Fig. 4.2 Complete tectonic 'stratigraphy' of the Semail (Oman ophiolite)

base of the ophiolite during initial detachment, and preserved in tectonic windows beneath the ophiolite. It consists of an inverted metamorphic sequence of strongly sheared   hornblende + plagioclase ± garnet ± clinopyroxene   granulite   and amphibolite facies gneisses overlying epidote amphibolites and a range of green-schist facies meta-cherts and meta-carbonates (Searle and Malpas 1980, 1982; Hacker et al. 1996; Gnos 1998; Cowan et al. 2014).

The mantle sequence includes mainly harzburgites and dunites and can reach a structural thickness of over 15 km (Fig. 4.3). Along the base of the mantle sequence layered harzburgites and dunites show an emplacement related ductile shear fabric (Fig. 4.4). Higher up-section dunites become more irregular and may represent melt flow pathways where the extraction of the basaltic component has left a depleted residual mantle dunite (Fig. 4.5). The mantle sequence hosts minor chromite deposits (Fig. 4.6) as well as some tonalitic–trondhjemitic dykes. In the north in UAE rare leucogranite dykes intrude the peridotites related to crustal melting of a sedimentary source after initial emplacement (see Chap. 7). The Moho, the geo-physical boundary between the mantle and the crust, is rarely an abrupt contact (Figs. 4.7 and 4.8); it is more commonly exposed as a 'Moho Transition zone' a complex zone of inter-layered harzburgites (olivine + orthopyroxene), dunites (olivine) and mafic gabbros (olivine + pyroxene + plagioclase) with large lenses of dunite (Fig. 4.9), exposed along the top of the mantle section (Boudier and Nicolas 1995). Late, dark-weathering wehrlite (olivine + clinopyroxene ± plagioclase) intrusions cut the Moho Transition zone and the lower crust gabbros, and are thought to be related to off-axis magmatism post-dating the early dykes and lavas and associated with subduction-related volcanism (Figs. 4.10 and 4.11). The lower oceanic crust is made of layered gabbros with cumulus textures related to cyclic batch melting. More primitive magmas are gabbro norites (olivine + orthopyrox-ene + clinopyroxene). The upper gabbros show large horizontal sill complexes that fed magma laterally away from an axial 'mush' zone. High-level intrusives include plagiogranites (plagioclase + hornblende ± quartz), the final fractionated phase of the melting process. Plagiogranites form the roof zone of the magma chamber where volatiles are concentrated. They also contain <1% zircon, a mineral used for U–Pb dating of ophiolite complexes. The gabbros merge upwards into the root zone of the sheeted dykes which transported doleritic-basaltic magma up to the pillow lavas. Some spectacular outcrops of dykes and sills of mixed gabbros (dark dioritic blocks enclosed in white felsic plagiogranite) termed 'vinaigette' intrude the gabbro sequence in the UAE. Many impressive outcrops in both Oman and UAE show 100% sheeted dykes, with chilled margins, indicative of a totally extensional tec-tonic setting, interpreted as an oceanic spreading center (Figs. 4.12 and 4.13). The sheeted dykes can be seen to feed the lower pillow lavas (Fig. 4.14).

Fig. 4.3 Mountain ranges composed entirely of harzburgite mantle sequence, east of the Semail Gap

Fig. 4.4 Banded ultramafic unit showing dunites (pale) and harzburgites (dark) along the base of the Muscat peridotite, hills behind Muttrah harbour

Fig. 4.5 Pale coloured dunite and darker harzburgite of Ibra block ophiolite

The volcanic section of the Oman ophiolite shows four major structural units. The lower Geotimes unit (V1) consists of classic pillow lavas (Fig. 4.15), overlain by up to 1100 m of Lasail unit (V2) island-arc tholeiites and boninites (high-Mg andesites) related to supra-subduction zone magmatism (Pearce et al. 1981; Alabaster et al. 1982; Lippard et al. 1986). Geochemistry of the volcanic sequence suggests that all lavas preserved in the Oman ophiolite are closer to supra-subduction zone magmas, including back-arc or fore-arc basalts (Pearce et al. 1981; McLeod et al. 2013). Red radiolarian cherts are sometimes interbanded within the pillow lavas or along the top of the ophiolite showing their origin in a deep ocean setting far from any continental clastic source. U–Pb zircon ages from the plagiogranites and gabbros constrain the age of the Oman–UAE ophiolite as Cenomanian, 96.12–95.50 million years old (Rioux et al. 2016). Ages of radiolaria from the interbedded cherts at the top of the ophiolite are also Cenomanian–Turonian in agreement with the zircon ages.

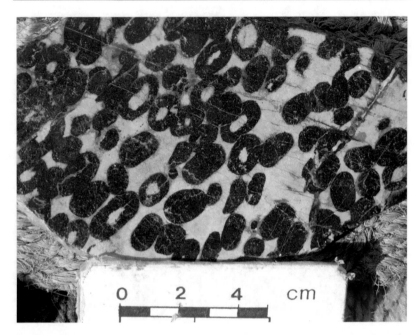

Fig. 4.6 Polished slab of 'polo' chromite showing black chromite and pale brown dunite, Wadi Jizzi region

The ophiolite section in UAE shows the complete ophiolite sequence up to the sheeted dykes exposed on the coast south of Fujairah. Although the Semail Ophiolite is also beautifully exposed in the south-eastern mountains from Ibra to Wadi Tayyin, the uppermost Lasail unit island arc tholeiites and boninites are missing in this region. The most complete profile through the ophiolite is the north Oman region around Wadi Jizzi, south to Wadi Salahi and Wadi Ahin. This region shows the full range of mantle and crustal rocks, including the spectacular Geotimes pillow lava outcrop in Wadi Jizzi, the Lasail and Aarja copper mineralisation in the V2 arc lavas, the beautiful 3D outcrops of 100% sheeted dykes in Wadi Salahi and Wadi Ahin, and the sharp Moho in Wadi Jizzi. This region is undoubtedly the best profile through any ophiolite sequence anywhere in the World, and it has been suggested that this region should be designated a UNESCO World Heritage Site and National GeoPark (Searle 2014).

Fig. 4.7 Hand specimen of the Semail Ophiolite Moho, separating dark upper mantle harzburgite below from pale coloured lower crust gabbro above

4.2 Metamorphic Sole

The most complete sections through the metamorphic sole of the Semail ophiolite occur in Wadi Tayyin in the eastern mountains, in the Al-Ajaiz and Fanjah regions near Muscat, in the Sumeini Window in northern Oman, and along the Masafi region of UAE (Searle and Malpas 1980, 1982; Hacker et al. 1996; Gnos 1998; Cowan et al. 2014). In all these areas, the highest level of the sole is composed of garnet + clinopyroxene + hornblende + plagioclase bearing granulites, or

Fig. 4.8 Sharp Moho contact between peridotites of the mantle sequence (right) and pale coloured layered gabbros of the lower crustal sequence (left), Fanjah–Nakhl road

amphibolites with granulite enclaves, immediately beneath the Semail thrust (Fig. 4.16). Thermobarometry indicates that the granulite enclaves formed at temperatures of 900–770 °C and pressures of 13–11 kbar, equivalent to depths of 35–40 km (Cowan et al. 2014). This depth is more than double that of the exposed total thickness of the Semail ophiolite meaning that basalts forming the amphibolite sole must have been subducted into the mantle at deeper levels that preserved in the ophiolite sequence. U–Pb zircon ages from partial melt pods within the amphibolites at Sumeini and Wadi Tayyin are 96.16–94.82 million years old, showing that this subduction zone must have underlain the Semail ophiolite at the same time as the crustal sequence was forming (Rioux et al. 2016).

Beneath the granulite and garnet + clinopyroxene amphibolites, garnet-free amphibolites are exposed and then epidote amphibolites showing an inverted metamorphic gradient. A range of metamorphosed cherts (banded quartzites)

Fig. 4.9 Layered gabbros (pale) and harzburgites (dark), Moho transition zone, Fanjah region

limestones (marbles) and rare volcanic tuffs form the greenschist facies part of the sole. Only in a few rare locations, as at Masafi, Sumeini, Asjudi, Al Ajaiz, Fanjah, and Wadi Tayyin, is the sole preserved in its original position, welded along the base of the ophiolite beneath the Semail thrust. Elsewhere it has been imbricated and in places broken up and incorporated as blocks in a tectonic mélange in the Haybi complex. The protolith rocks of the metamorphic sole are all components of the underlying Haybi complex, tholeiitic and alkali basalts, Triassic Exotic limestones, cherts and rare pelites. The protoliths are not the same as the ophiolite complex. Thus, it is most likely that during, or soon after, subduction initiation, old cold oceanic crust comprising Triassic–Jurassic ocean crust with Late Triassic exotic limestones and deep-ocean cherts was subducted to the NE beneath a supra-subduction zone spreading center along which the Semail ophiolite was forming crust at the same time. The extrusion of the sole and accretion along the base of the ophiolite initiated the obduction process at ∼96–95 million years ago.

Fig. 4.10 Moho Transition zone showing black coloured wehrlite intruding gabbros and peridotites, Wadi Fanjah

4.3 Ophiolite Origin and Obduction

Earlier studies on the Oman ophiolite assumed that the sequence was formed along a Mid-Ocean Ridge (Reinhardt 1969; Coleman 1981; Nicolas et al. 1996) and recently comparisons have been made with fast-spreading mid-ocean ridges such as the East Pacific Rise (Boudier et al. 1997). Detailed field and geochemical studies on the more complete volcanic sequence in the northern Oman Mountains led to the discovery of island arc tholeiitic lavas and boninites in the upper pillow lavas, and led to the proposal that the Semail ophiolite formed as an immature island arc sequence or more accurately in a supra-subduction zone tectonic setting (Pearce et al. 1981; Alabaster et al. 1982; Ishikawa et al. 2002; Kusano et al. 2014). Several other facts support the supra-subduction zone model. Firstly, the simultaneous formation of the metamorphic sole granulites and amphibolites accreted to the base of the peridotite (Searle and Malpas 1980, 1982; Cowan et al. 2014) and

Fig. 4.11 Complex relationships of gabbros intruding peridotites, Moho Transition zone

formed at 35–40 km depth in a NE-dipping subduction zone were formed at the same time as the overlying crustal sequence (Rioux et al. 2016). Secondly, large chromitite pods, abundant along the Oman ophiolite, have not been reported from modern mid-ocean ridges, either by drilling or dredging (Rollinson 2014). Thirdly, detailed geochemistry of all ophiolite lavas showed that even the Geotimes (V1) pillow lavas were geochemically distinct from MORB with higher magmatic water contents (McLeod et al. 2013). Finally, numerous leucogranite dykes were found intruding the mantle sequence of the northern ophiolite in UAE. These potassic granites are highly aluminous containing andalusite, garnet, tourmaline and mus-covite and clearly prove that they formed by crustal melting of a pelitic (shaley) protolith during obduction of the ophiolite over the continental margin (Cox et al. 1999: Rollinson 2015).

The obduction of the Semail ophiolite began during formation of the ophiolite in the Cenomanian and emplacement lasted approximately 20 million years

Fig. 4.12 Outcrop photo of sheeted dykes showing chilled margins, Fujairah, UAE

Fig. 4.13 Sheeted dyke complex, Wadi Ahin, representing 100% oceanic crustal extension along a spreading ridge

Fig. 4.14 Sheeted dykes feeding lower Geotimes unit pillow lavas, Wadi Ahin

Fig. 4.15 Geotimes (V1) unit pillow lavas, Wadi Jizzi

Fig. 4.16 Semail thrust with mantle harzburgites thrust over garnet + diopside amphibolite of the metamorphic sole, Sumeini Window, north Oman

(from ~95 to 75 million years ago) evolving from a deep, ductile deforming subduction zone into a thin-skinned brittle fold and thrust belt with time. In the Bani Hamid thrust sheet in the northern mountains around Madhah and UAE tightly folded granulite facies cherts and carbonates represent stacked-up deep crustal thrust sheets (Searle et al. 2014). During the later stages of obduction the continental margin of Arabia was subducted to depths of 80–100 km as evidenced from the eclogites at As Sifah, formed at 20–23 kbar and temperatures of 540 °C (Searle et al. 1994, 2004). These eclogites formed 79 million years ago (Warren et al. 2003), 15 million years after formation of the metamorphic sole and ophiolite. The exhumation of the eclogites, blueschists and carpholite-bearing meta-sedimentary rocks of the Muscat–Sifah region was accompanied by intense, isoclinal folding on all scales, sheath folding as seen along Wadi Mayh and ductile fabrics that indicate exhumation toward the SSW beneath passive roof faults. These top-to-NNE (base to SSW) fabrics have been mistaken for representing a phase of extension following high-pressure metamorphism.

4.4 Masirah Ophiolite

The Masirah ophiolite exposed on the island of Masirah off the southeast coast of Oman is a different ophiolite from the Semail ophiolite along the northern Oman Mountains (see Chap. 14). Masirah has both a mantle and crustal sequence, but the crust is only about 2–2.5 km thick, as opposed to 6–8 km in the Semail ophiolite. The gabbro sequence is highly condensed (only about 500 m thick) compared to the north Oman section. U–Pb zircon age dating suggests a Jurassic age of formation at 156–150 million years ago (Peters et al. 1997). The basaltic lavas are 250–500 m thick, have a MORB geochemistry and contain plagioclase, olivine, clinopyroxene and Cr-spinel phenocrysts (Peters et al. 1997; Peters and Mercolli 1998). A sequence of ocean island type alkali basalts of Cretaceous age (130–125 million years old) overlie the Masirah ophiolite (Meyer et al. 1996). Rollinson (2017) suggested that the Masirah ophiolite was a part of the Indian ocean crust, transpressionally uplifted by two strike-slip faults that occur offshore along both the west and east coasts of the island.

Magnetic, seismic and structural data suggest that the Masirah ophiolite is part of an elongate 'basement' ridge of ophiolite that extends along the SE coastal margin of Oman (Mountain and Prell 1990). This ophiolite belt, together with two other slices of mantle rocks exposed at Ras Jibsch and Ras Madrakah, and the Batain mélange, was emplaced towards the WNW during the late Maastrichtian–Early Palaeocene (Shackleton and Ries 1990; Shackleton et al. 1990; Peters et al. 1997; Schreurs and Immenhauser 1999). This timing slightly post-dates the final stages of emplacement of the North Oman Semail ophiolite belt (ca. 95–67 Ma), although the earlier stages of obduction of the Masirah ophiolite remain unknown, as the base is not exposed on land.

4.5 Late Cretaceous Folding and Thrusting

The major orogenic period in the North Oman–UAE Mountains was the Late Cretaceous period (Cenomanian to Maastrichtian) when the Semail Ophiolite and underlying thrust sheets (Haybi and Hawasina complexes) were emplaced from NE to SW onto the drowned continental margin of Arabia. This emplacement occurred almost entirely below sea-level, with the mountains just breaching the surface since the latter part of the orogeny. Thin laterite (paleo-soil) horizons occur on the eroded surface of the ophiolite in several places. The thrusting effectively shortened an ocean width of over 200 km from reconstruction of the folds and thrusts in the Hawasina and Haybi complexes. The metamorphic profile in the sole is inverted with higher temperature-pressure rocks structurally above lower grade rocks. The heat for metamorphism must therefore come from the mantle wedge beneath the preserved ophiolite.

Subduction initiation ∼96 million years ago resulted in deep level ophiolite obduction processes to commence with exhumation of the metamorphic sole and its accretion along the base of the ophiolite. Thrusting processes evolved from subduction-exhumation ductile shearing processes to thin-skinned fold and thrust belt tectonics with time. Thrusting mainly propagated towards the continent with time, with the sequence of thrusting going from Semail ophiolite to the Haybi, Hawasina and Sumeini thrust sheets with time. Some late stage out-of-sequence thrusts have been recorded at outcrop scale in the mountains. The youngest thrusts are the far southwest foreland jebels (Jebel Salakh arch) where they extend down into the shelf carbonates beneath the Hawasina thrust sheets. It is possible that the latest phase of Late Cretaceous thrusting extended right down to basement levels with the culmination of the giant Jebel Akhdar and Saih Hatat domes.

After the main Hawasina, Haybi and Semail ophiolite thrust sheets were emplaced across the continental margin the leading margin of the continent was pulled down the subduction zone. This is known from the high-pressure eclogites at As Sifah and the overlying blueschist and carpholite-bearing meta-sediments. All these units were metamorphosed at high-pressure, 79 million years ago, some 15 million years after subduction initiation and the beginning of ophiolite obduction and thrusting.

4.6 Continental Subduction and High-Pressure Metamorphism

High-pressure eclogite and blueschist facies rocks are found only along the northern part of the mountains between the Ruwi valley near Muscat southeastwards to As Sifah and Quriat (Fig. 4.17; see Chap. 11). The structurally deepest units are seen in As Sifah where boudins of meta-volcanic eclogites occur within strongly deformed carbonates and schists thought to be metamorphosed Permian Saiq Formation and metamorphosed Amdeh Formation sandstones and quartzites (Fig. 4.18). The eclogites are impressive dark weathering rocks showing large reddish garnets, dark blue glaucophane amphiboles, green pyroxenes and shiny white phengite mica. The eclogites and the surrounding calc-schists have impressive S-C fabrics showing top-NNE motion, related to SSW-extrusion of footwall high-pressure rocks rather than any true extension. Carbonates are far more buoyant than basalts or eclogites so

Fig. 4.17 Digital Elevation model of northern Saih Hatat, view to SW. As Sifah is at bottom left and Wadi Mayh is the prominant wadi in the center. The drowned coastline of Bandar Jissah and Bandar Khyran region is seen in bottom right (north)

cannot be easily subducted. It is likely that the continental margin was pulled down the subduction zone, and then after the oceanic basaltic slab broke off and sank into the mantle, the carbonates with their eclogite boudins were left hanging at depth in the surrounding mantle. Buoyancy forces resulted in their rapid exhumation, back along the same subduction zone. Intense isoclinal folding is related to the exhumation of these high-pressure rocks (Fig. 4.18).

In northern Saih Hatat a series of ductile shear zones, dipping towards the north or NE have lower pressure rocks above and higher pressure rocks below. Deformation is intense with isoclinal folding in the Permian shelf carbonates widespread (Fig. 4.19). The largest of the ductile shear zones are the As Sifah, As Sheik, Al Khuryan and Yenkit shear zones (Searle et al. 2004). Each of these has isoclinal folds, and spectacular ductile shear fabrics concentrated along zones of high strain. These shear zones all formed between 79 million years ago (peak eclogite metamorphism) and about 68 million years ago. The blueschist facies rocks above the As Sifah eclogites are characterised by lawsonite pyroxene in meta-volcanic units and the mineral carpholite in meta-sedimentary units, both of which are indicative

Fig. 4.18 The dark coloured As Sifah eclogites at beach level surrounded by pale coloured calc-schists, southeast of Muscat, location of the high-pressure rocks

of high pressures. The high-pressure rocks along northern Saih Hatat extend down south as far as the basement. Along Wadi Mayh and Wadi Aday spectacular sheath folds can be seen in carbonates of the Permian and Mesozoic sequence. These sheath folds have been mapped out in great detail (Cornish and Searle 2017), and the progression from undeformed and unmetamorphosed shelf sequences around the southern and western margin of Saih Hatat to highly folded and sheared rocks is magnificently exposed in the northwest part of Saih Hatat and south of the Ruwi valley. The eclogites, and the giant Wadi Mayh sheath fold structures are all of international significance and geologically extremely important, so it is imperative that these sites are preserved as GeoParks.

4.7   Musandam Culmination

The culmination and uplift of Musandam was a post-Eocene event thought to have been related to the first collisional effects of the Arabia–Iran plate collision (see Chap. 6). The Permian–Mesozoic shelf carbonates have been thrust above the stable

Fig. 4.19 Isoclinal folds in Permian Saiq Formation limestones, mountains above Wadi Aday

foreland shelf along the Hagab thrust, a west-vergent thrust that displays a huge recumbent anticline along the western margin of the Musandam in Ras al Khaimah. This thrust cuts up-section as far as the Miocene in the footwall as seen in offshore seismic lines. Several more thrusts have been imaged on seismic lines offshore Ras al Khaimah cutting down into the pre-Permian basement rocks. The northern Musandam shows that the anticline axis plunges gently north towards the Straits of Hormuz and the syntaxis where the orogenic trend swings around to align with the WNW-ESE trend of the Zagros mountains in Iran. Along the southeast margin of Musandam in UAE a large-scale normal fault, the Dibba fault, runs from the nose of Musandam along Batha Mahani to Dibba and trends NNE along the east coast. The oldest Permian shelf carbonates are exposed along the east coast cliffs north of Dibba, implying that the whole Musandam peninsula is geologically tilted to the west, and uplifted along the east. In contrast, the present-day coastline shows spectacular drowned wadi systems of Musandam and deep fjords which were formed by recent north and eastward tilting of the whole Musandam peninsula.

4.8   Cenozoic Deformation

Following the Late Cretaceous obduction of the Semail ophiolite and thrusting of underlying Haybi and Hawasina complex thrust sheets onto the Arabian margin, the ending of this phase of orogeny is marked by a regional unconformity, above which shallow marine rudist-bearing limestones (Simsima Formation) were deposited over all underlying units. Death assemblage beds composed almost entirely of rudists and gastropods are exposed in Jebel Sumeini and across the northern Oman and UAE mountains. Above this, highly fossiliferous Paleocene–Eocene limestones record stable shallow marine conditions across the Oman–UAE area for at least 20 million years. After the Oligocene, another phase of mountain building started as the mountains began to uplift. Post-Eocene, possibly as young as Miocene, folding is seen along the SW flank of the mountains notably in Jebel Hafit and the foreland jebels of UAE and Oman and also along the north coast region from Ras al Hamra–Qurum areas southeast to Sur. These folds are large-scale structures, generally showing box-fold geometry and prominent kink bands with minor thrusting. In the Ras al Hamra–Qurum and Bandar Jissah areas near Muscat, the folds have NNW-SSE aligned axes, a trend that is almost at right angles to the Late Cretaceous structures in the high-pressure region of northern Saih Hatat. In the Abat basin south of Sur, rocks as young as Late Miocene are affected by large scale folding indicting that the age of this deformation event could be as young as Pliocene–Pleistocene.

The Late-Oligocene–Miocene structures associated with the earliest collision of Arabia and Iran as seen in the Musandam peninsula can be traced north to a point in the Straits of Hormuz where the east-west trending folds and thrusts of the Late Cenozoic Zagros fold belt meet (Fig. 4.20). Thus Musandam marks a transition between the dominantly Late Cretaceous ophiolite obduction related structures in the Oman Mountains to the south, and the dominanly continental collision rela-ted Late Cenozoic structures in the Zagros fold-thrust belt to the northwest.

4.9   Regional Tectonics

Two major orogenic events dominate the stress field in Oman during the Late Cretaceous, firstly the emplacement of the Semail ophiolite and underlying thrusts sheets from NE to SW onto the passive margin of Arabia, and secondly the emplacement of the Masirah ophiolite and underlying Batain mélange from ESE to WNW along the southeast Oman coast. The resultant bi-axial compressive stress

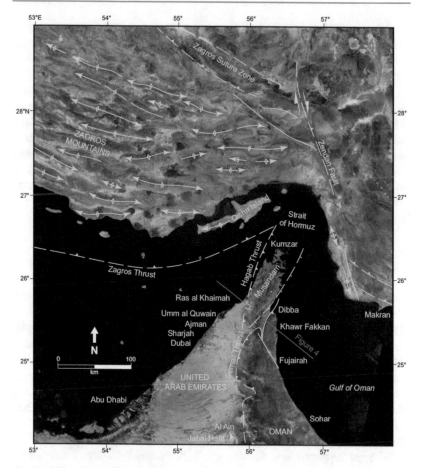

Fig. 4.20 Landsat satellite map of the Musandam peninsula, Straits of Hormuz syntaxis and Zagros fold and thrust belt, SW Iran

field resulted in a dome and basin type fold pattern along the North Oman Mountains (Fig. 3.19). The major anticline axis swings around from NW-SE in the Hawasina Window to ENE-WSW along Jebel Akhdar, then swings around through 90° along the Jebel Nakhl NNE-SSW trend before swinging back to WNW-ESE along Saih Hatat. Four main NNE-SSW aligned anticline axes at right angles to the main trend are apparent from the map pattern: (1) Wadi Bani Umar–central Hawasina; (2) Rustaq to Bahla; (3) Saiq to Jebel Nakhl; and (4) Hubat–Baid–Ras al

Hamra (Fig. 3.19). Where two anticline axes meet, a structural dome is present (Hawasina Window, Jebel Akhdar, Saih Hatat). The ophiolite blocks of Rustaq, Bahla and Semail are located in structural depressions where two syncline axes intersect, in-between the shelf carbonate domes.

The bi-axial fold interference pattern is undoubtedly late stage because it affects all the allochthonous thrust sheets after their emplacement onto the shelf margin. Mid-Cenozoic fold axes orientations are parallel to the Late Cretaceous fold axes indicating two phases of NE-SW compression. Folds in the Cenozoic jebels along the southwestern flank of the mountains (e.g.: Fahud, Natih, Ibri, Dhank structures; Fig. 3.19) have a similar orientation to the Late Cretaceous structures along the mountains. West of Muscat in the Ras al Hamra area, fold axes in Paleogene rocks are aligned NNE-SSW parallel to the Late Cretaceous sheath folds in the Saih Hatat Window.

The fracture pattern in the Oman foreland south and southwest of the mountains is very different from the Oman Mountains (see Chap. 8). The Late Cretaceous foreland fracture pattern, interpreted from 3D seismic data as the result of NW-SE to WNW-ESE compressive stress (Filbrandt et al. 2006) is also very consistent across nearly 1000 km from the Arabian Gulf to the Indian Ocean. South of the Hawasina basal thrust in the foreland there is very little evidence of S- or SW-directed thrusts or fractures associated with the Oman Mountains trend (Filbrandt et al. 2006). In the foreland, SW of the Oman Mountains, from Lekhwair to the Ghaba salt basin, two dominant fault sets are oriented WNW-ESE and NNW-SSE (Filbrandt et al. 2006). The highest fault density occurs in the competent Mesozoic shelf carbonate unit, whereas the clastic dominated Aruma Group absorbs strain by buckling and internal structural thickening. Several prominent faults, such as the Maradi and Burhan faults, developed during the Campanian and form en-echelon faults showing sinistral shear.

Many of the north Oman Shuaiba and Natih Formation oil traps formed during the Campanian and are associated with large steep faults that were active during deposition of the Aruma foredeep sediments (Turonian to Campanian). Thickening of the Fiqa Formation shales towards the SW-dipping Fahud normal fault formed the Fahud structure. The Fahud anticline affects the Fiqa Formation and overlying Paleogene limestones. Post-Eocene contraction caused local inversion of earlier faults such as the Fahud fault and reversal of slip along the Maradi fault, which shows a dextral sense of shear today. These faults may have provided vertical conduits for migration of hydrocarbons from deeper reservoirs (for example, Gharif Formation) up into shallower reservoirs (for example, Shuaiba, Natih Formations).

Filbrandt et al. (2006) suggested that the NW-SE to WNW-ESE maximum horizontal stress present along the Oman foreland was a result of an oblique collision of the Indian continent with the Arabian Plate during the Santonian–Campanian. Although there is certainly a fracture pattern that is consistent with this stress field, the fractures have very small, or no offsets, and several faults have been re-activated during an evolving stress regime. The deformation in the foreland and along the SE coast of Oman is not compatible with a continental collision such as presently seen around the northern margins of the Indian Plate (hundreds of kilometers of shortening, regional metamorphism, etc.). Instead, it is more likely that the fracture pattern in the Oman foreland is related to WNW-directed emplacement of the Masirah ophiolite and Batain mélange.

It is possible that the NNE-SSW trending fold axes in the North Oman Mountains were related to the Masirah ophiolite emplacement event. In this case, the biaxial fold interference pattern in Jebel Akhdar–Saih Hatat could have been the result of refolding of the earlier NW-SE aligned folds (Jebel Akhdar trend) by a slightly later set of NNE-SSW aligned folds (Jebel Nakhl–Semail Gap trend).

4.10 Jebel Akhdar and Saih Hatat Culmination

The Jebel Akhdar culmination shows a single very large-scale 30–35 km half-wavelength anticline with a flat top (Fig. 4.21). Dips increase outward from the axis such that the style is one of a gentle rounded box fold. Along the southern flank dips are gentle (15–20° S) but increase towards the NW-plunging nose where a distinct overturning, verging SW is apparent. This may be a kink band that progressively developed into a steep thrust from Jebel Akhdar towards the Hawasina Window at depth. Along the northern flank dips are 25–35° N. The upper contact of the shelf carbonates is presently a listric normal fault along the northern flank of Jebel Akhdar, and the west and east flanks of Jebel Nakhl are also late normal faults related to the uplift of the footwall shelf carbonates. Ophiolite mantle sequence rocks rest directly on the Wasia Formation top shelf carbonates with most of the Haybi and Hawasina thrust sheets faulted down. Throw on these late-stage normal faults could be as much as 4–5 km.

Continuity of Hawasina, Haybi and Semail ophiolite thrust sheets over Jebel Akhdar prior to folding is apparent from the presence of ophiolite all around its south, west and northern flanks (Bahla–Sint–Hawasina–Rustaq ophiolite blocks) and south and eastern flank (Nizwa–Semail ophiolite blocks). West of the Wadi Nakhr 'Grand Canyon' and Jebel Shams summit, Haybi thrust slices comprising Triassic alkali volcanics and Oman exotics (Jebels Kawr, Misfah and un-named

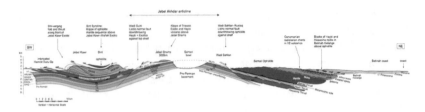

Fig. 4.21 Cross-section across Jebel Akhdar, after Searle (2007; GeoArabia)

exotics along the crest of the fold) are more-or-less continuous over the south-western limb of the Jebel Akhdar fold.

The large scale of the Jebel Akhdar fold resulted in extensive bedding plane slip during flexure. Along the south flank of Jebel Akhdar, little or no internal folding or thrusting has been observed in many of the deeply incised wadi sections (e.g.: Wadis Nakhr, Tanuf, Mu'aydeen). Some small-scale intra-formational thickening occurred by internal duplexing locally along the north flank of Jebel Akhdar (Breton et al. 2004; Fig. 7). These are very small-scale folds and thrusts, restricted to the Cretaceous part of the shelf sequence and are not continuous up to the underlying shelf or overlying allochthonous units. These duplexes are not regional north-verging folds (Gray and Gregory 2000), but local features related to thickening along the flanks of the main Jebel Akhdar fold. Similar internal thrusts and folds in the shelf sequence of northern Saih Hatat are south-vergent (for example Yiti duplex; Searle 2007). The north-vergent backfolds of the Hawasina Window are much larger scale, late-stage structures affecting the overlying Semail ophiolite thrust sheet as well.

Bouger gravity data suggest a crustal thickness beneath Jebel Akhdar of 47 km (Ravaut et al. 1997), or as much as 48–51 km (Al-Lazki et al. 2002). This thick crustal root is surprising, being only 75 km inland from the present-day coastline. Even beneath the Batinah coast the crustal thickness is as much as 39–42 km (Al-Lazki et al. 2002). Because of this, it is suggested that the Arabian continental crust must have extended a long way offshore buried beneath the thick Ceno-zoic sedimentary sequence. This would increase the distance of obduction of the Semail ophiolite and underlying thrust sheets even more than proposed from surface geology.

The Saih Hatat culmination (Fig. 4.22) is more complex with a relatively simple southern flank where the Mesozoic shelf carbonate stratigraphy is intact, but a much more complex northern flank. Northern Saih Hatat shows increasingly higher pressure metamorphism with carpholite schists, blueschists and eclogites all

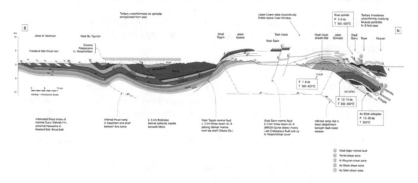

Fig. 4.22 Cross-section of Saih Hatat culmination, after Searle (2007, GeoArabia)

associated with extreme deformation, including isoclinal folding, high-strain fabrics and ductile shear zones. North-dipping normal faults and ductile shear zones are associated with exhumation of high-pressure rocks from the subduction along the footwall. Two possible models for the structure of Jebel Akhdar have been proposed. The first model is a simple fold with no thrusting at depth (Glennie et al. 1973, 1974; Al-Lazki et al. 2002). The second model shows the anticline as a thrust culmination, either with a flat-lying detachment at depth ramping up in front of the Jebel Salakh frontal fold (Searle 1985; Fig. 11; Bernoulli and Weissart 1987; Fig. 2; Hanna 1990; Fig. 11; Cawood et al. 1990, Fig. 11), or with steeper basement faults (Mount et al. 1998). Although the dips of the Jebel Akhdar range are relatively gentle there is a SW-verging asymmetry to the fold towards the west (Fig. 4.21). Geometrical constraints suggest that a deep flat within the pre-Permian basement is likely albeit with little horizontal translation. Although the Jebel Akhdar culmination must have been a positive feature during the latest stages of Late Cretaceous deformation, the main period of growth was probably Oligocene–Early Miocene. Fission track apatite data suggests that Late Eocene–Oligocene uplift and erosion was followed by slower cooling in the Neogene (Hansman et al. 2017).

References

Alabaster T, Pearce J, Malpas J (1982) The volcanic stratigraphy and petrogenesis of the Oman Ophiolite complex. Contrib Miner Petrol 81(3):168–183

Al-Lazki AI, Seber D, Sandvol E, Barazangi M (2002) A crustal transect across the Oman mountains on the eastern margin of Arabia. GeoArabia 7:47–78

Ambrose TK, Searle MP (2019) 3-D structure of the Northern Oman–UAE Ophiolite: Widepsread, Short-lived, Suprasubduction zone magmatism. Tectonics 38. https://doi.org/10.1029/2018TC005038

Ambrose TK, Wallis D, Hansen LN, Waters DJ, Searle MP (2018) Controls on the rheological properties of peridotite at a paleosubduction interface: a transect across the base of the Oman-UAE ophiolite. Earth Planet Sci Lett 491:193–206

Bernoulli D, Weissart H (1987) The upper Hawasina nappes in the central Oman mountains stratigraphy, palinspastics and sequence of nappe emplacement. Geodinamica Acta 1:47–58

Boudier F, Nicolas A (1995) Nature of the Moho transition zone in the Oman Ophiolite. J Petrol 36(3):777–796

Boudier F, Nicolas A, Ildefonse B, Jousselin D (1997) EPR microplates, a model for the Oman ophiolite. Terra Nova 9(2):79–82

Breton JP, Béchennec F, Le Métour J, Moen-Maurel L, Razin P (2004) Eoalpine (Cretaceous) evolution of the Oman Tethyan continental margin: insights from a structural field study in Jebel Akhdar (Oman mountain). GeoArabia 9:41–58

Coleman RG (1981) Tectonic setting for Ophiolite Obduction in Oman. J Geophys Res 86:2497–2508

Cowan RJ, Searle MP, Waters DJ (2014) Structure of the Metamorphic Sole to the Oman Ophiolite, Sumeini Window and Wadi Tayyin: implications for ophiolite obduction processes. In: Rollinson HR, Searle MP, Abbasi IA, Al-Lazki A, Al-Kindy MH (eds) Tectonic evolution of the Oman Mountains. Geological Society, London, Special Publication, vol 392, pp 155–176

Cox JS, Searle MP, Pedersen RB (1999) The petrogenesis of leucogranite dykes intruding the northern Semail ophiolite, United Arab Emirates: field relations, geochemistry and Sr/Nd isotope systematics. Contrib Miner Petrol 137:267–287

Filbrandt JB, Al-Dhahab S, Al-Habsy A, Harris K, Keating J, Al-Mahruqi S, Ozkaya I, Richard PD, Robertson T (2006) Kinematic interpretation and structural evolution of North Oman, Block 6, since the Late Cretaceous and implications for timing of hydrocarbon migration into Cretaceous reservoirs. GeoArabia 11(1):97–139

Glennie KW, Boeuf MG, Hughes-Clarke MHW, Moody-Stuart M, Pilaar WF, Reinhardt BM (1973) Late Cretaceous nappes in the Oman Mountains and their geologic evolution. Bull Am Assoc Pet Geol 57:5–27

Glennie KW, Boeuf MG, Hughes-Clarke MHW, Moody-Stuart M, Pilaar WF, Reinhardt BM (1974) Geology of the Oman Mountains. Verhandelingen Koninklijk Nederlands geologisch mijnbouwkundidg Genootschap no 31, 423 p

Gnos E (1998) Peak metamorphic conditions in garnet amphibolites beneath the Semail ophiolite: implications for an inverted pressure gradient. Int Geol Rev 40:281–304

Gray DR, Gregory RT (2000) Implications of the structure of the Wadi Tayyin metamophic sole, the Ibra-Dasir block of the Semail ophiolite, and the Saih Hatat window for late stage extensional ophiolite emplacement. Marine Geol Res 21:211–227

Hacker BR, Mosenfelder JL, Gnos E (1996) Rapid emplacement of the Oman ophiolite: thermal geochronological constraints. Tectonics 15:1230–1247

Hanna SS (1990) The Alpine deformation of the central Oman mountains. In: Robertson AHF, Searle MP, Ries AC (eds) The geology and tectonics of the Oman region. Geological Society, London Special Publication 49, pp 341–359

Hansman RJ, Ring U, Thomson S, den Brok B, Stübner K (2017) Late Eocene uplift of the Al Hajar mountains, Oman, supported by stratigraphy and low-temperature thermochronology. Tectonics 36:3081–3109

Hopson CA, Coleman RG, Gregory RT, Pallister JS, Bailey EH (1981) Geologic section through the Samail Ophiolite and associated rocks along a Muscat–Ibra transect, Southeastern Oman Mountains. J Geophys Res 86:2527–2544

Ishikawa T, Nagaishi K, Umino S (2002) Boninitic volcanism in the Oman ophiolite: implications for thermal condition during transition from spreading ridge to arc. Geology 30:899–902

Kusano Y, Hayashi M, Adachi Y, Umino S, Miyashita S (2014) Evolution of volcanism and magmatism during initial arc stage: constraints on the tectonic setting of the Oman ophiolite. In: Rollinson HR, Searle MP, Abbasi IA, Al-Lazki A, Al-Kindy MH (eds) Tectonic evolution of the Oman Mountains. Geological Society, London, Special Publication, vol 392, pp 177–193

Lippard SJ, Shelton AW, Gass IG (1986) The Ophiolite of Northern Oman. Blackwell Scientific Publications, Oxford, 178 p

McLeod C, Lissenberg J, Bibby LE (2013) 'Moist MORB' axial magmatism in the Oman ophiolite: the evidence against a mid-ocean ridge origin. Geology 41:459–462

Meyer J, Mercolli I, Immenhauser A (1996) Off-ridge alkaline magmatism and seamount volcanoes in the Masirah island ophiolite, Oman. Tectonophysics 267:187–208

Mount VS, Crawford RIS, Bergman SC (1998) Regional structural style of the central and southern Oman mountains: Jebel Akhdar, Saih Hatat and the Northern Ghaba basin. GeoArabia 3:475–490

Mountain GS, Prell WL (1990) In: Robertson AFH, Searle MP, Ries AC (eds) The Geology and Tectonics of the Oman Region. Geological Society, London, Special Publication no 49, pp 725–744

Nicolas A, Boudier F (1995) Mapping oceanic ridge segments in Oman ophiolite. J Geophys Res 100(B):6179–6197

Nicolas A, Boudier F, Idelfonse B (1996) Variable crustal thickness in the Oman Ophiolite: implications for oceanic crust. J Geophys Res 101(B):17,941–17,950

Pearce JA, Alabaster T, Shelton AW, Searle MP (1981) The Oman ophiolite as a Cretaceous arc-basin complex: evidence and implications. Philos Trans R Soc Lond A300:299–317

Peters Tj, Immenhauser A, Mercolli I, Meyer J (1997) Geological map of Masirah north and Masirah south: explanatory notes, scale 1:50,000. Ministry of Petroleum and Minerals, Muscat, Oman

Peters Tj, Mercolli I (1998) Extremely thin oceanic crust in the Proto-Indian Ocean: evidence from the Masirah Ophiolite, Sultanate of Oman. J Geophys Res 193:677–689

Ravaut P, Bayer R, Hassani R, Rousett D, Al Yaya'ey A (1997) Structure and evolution of the northern Oman margin: gravity and seismic constraints over the Zagros-Makran-Oman collision zone. Tectonophysics 279:253–280

Reinhardt BM (1969) On the genesis and emplacement of ophiolites in the Oman Mountains geosyncline. Schweiz Minerl Petrogr Mitt 49:1–30

Rioux M, Bowring S, Kelemen P, Gordon S, Miller R, Dudas F (2013) Tectonic development of the Samali ophiolite: high-precision U–Pb zircon geochronology and Sm–Nd isotopic constraints on crustal growth and emplacement. J Geophys Res 118:1–17

Rioux M, Garber J, Bauer A, Bowring S, Searle MP, Keleman P, Hacker B (2016) Synchronous formation of the metamorphic sole and igneous crust of the Semail ophiolite: new constraints on the tectonic evolution during ophiolite formation from high-precision U–Pb zircon geochronology. Earth Planet Sci Lett 451:185–195

Rollinson H (2014) Plagiogranites from the mantle section of the Oman Ophiolite. In: Rollinson HR, Searle MP, Abbasi IA, Al-Lazki A, Al-Kindy MH (eds) Tectonic evolution of the Oman Mountains. Geological Society, London, Special Publication vol 392, pp 247–262

Rollinson H (2015) Slab and sediment melting during subduction initiation: Granitoid dykes from the mantle section of the Oman ophiolite. Contrib Miner Petrol 170:1177–1179

Rollinson H (2017) Masirah—the other Oman ophiolite: a better analogue for mid-ocean ridge processes? Geosci Front 8:1253–1262

Schreurs G, Immenhauser A (1999) West-northwest directed obduction of the Batain Group on the eastern Oman continental margin at the Cretaceous–Tertiary boundary. Tectonics 18:148–160

Searle MP (1985) Sequence of thrusting and origin of culminations in the northern and central Oman mountains. J Struct Geol 7:129–143

Searle MP (2007) Structural geometry, style and timing of deformation in the Hawasina window, Al Jabal al Akhdar and Saih Hatat culmionations, Oman mountains. GeoArabia 12:99–130

Searle MP (2014) Preserving Oman's geological heritage: proposal for establishment of World Heritage Sites, National GeoParks and Sites of Special Scientific Interest (SSSI). In: Rollinson HR, Searle MP, Abbasi IA, Al-Lazki A, Al-Kindy MH (eds) Tectonic evolution of the Oman Mountains. Geological Society, London, Special Publication vol 392, pp 9–44

Searle MP, Cox JS (1999) Tectonic setting, origin and obduction of the Oman ophiolite. Geol Soc Am Bull 111:104–122

Searle MP, Malpas J (1980) The structure and metamorphism of rocks beneath the Semail Ophiolite of Oman and their significance in ophiolite obduction. Trans R Soc Edinb Earth Sci 71:247–262

Searle MP, Malpas J (1982) Petrochemistry and origin of sub-ophiolite metamorphic and related rocks in the Oman Mountains. J Geol Soc Lond 139:235–248

Searle MP, Warren CJ, Waters DJ, Parrish RR (2004) Structural evolution, metamorphism and restoration of the Arabian continental margin, Saih Hatat region, Oman Mountains. J Struct Geol 26:451–473

Searle MP, Waters DJ, Martin HN, Rex DC (1994) Structure and metamorphism of blueschist–eclogite facies rocks from the northeastern Oman Mountains. J Geol Soc Lond 151:555–576

Searle MP, Cherry AG, Ali MY, Cooper DJW (2014) Tectonics of the Musandam Peninsula and northern Oman Mountains: from ophiolite obduction to continental collision. GeoArabia 19(2):135–174

Shackleton RM, Ries AC (1990) Tectonics of the Masirah Fault zone and eastern Oman. In: Robertson AFH, Searle MP, Ries AC (eds) The Geology and Tectonics of the Oman Region. Geological Society, London, Special Publication no 49, pp 715–724

Shackleton RM, Ries AC, Bird PR, Filbrandt JB, Lee CW, Cunningham GC (1990) The Batain Melange of NE Oman. In: Robertson AFH, Searle MP, Ries AC (eds) The Geology and Tectonics of the Oman Region. Geological Society, London, Special Publication no 49, pp 673–696

Warren CJ, Parrish RR, Searle MP, Waters DJ (2003) Dating the subduction of the Arabian continental margin beneath the Semail ophiolite. Geology 31:889–892

Oil and Gas

5

The oil-rich Arabian (Persian) Gulf region contains nearly half the World's oil and gas reserves. This zone stretches from Iraqi Kurdistan and the Zagros mountains of Iran through the Arabian Gulf region to Saudi Arabia, Kuwait, Qatar, the United Arab Emirates and Oman. Following the Late Carboniferous–Permian break-up of the supercontinent Gondwana about 300 million years ago, the northern margin of Arabia was a long-lasting passive continental margin, slowly accumulating biogenic rich sediments through time in warm, relatively shallow waters. More than 150 million years of stable sedimentation was abruptly ended approximately 95 million years ago by an orogenic mountain building event when a massive thrust slice of Tethyan ocean crust and upper mantle (the Semail Ophiolite) was thrust from NE to SW onto the Arabian continental plate margin. These ophiolites are perfectly preserved in the Oman–UAE Mountains but similar rocks, albeit highly deformed and poorly preserved, occur all along the Zagros suture in Iran. Remnant ophiolites of similar Late Cretaceous age as the Oman ophiolite are preserved at Neyriz and Kermanshah along the Zagros suture zone, along with old Tethyan oceanic sediments and mélanges. This ophiolite emplacement event preceded the collision of the continental part of the Arabian plate with the central Iran plate and closure of the Tethyan ocean approximately twenty million years ago. The collision of the Arabian plate with central Iran resulted in the Zagros orogeny and this event formed many of the anticlinal oil and gas traps along the Gulf region.

The Middle East is known for its enormous reservoirs of oil and gas. The first clues to this hidden reserve come in the form of natural oil and gas seeps when tar sands are formed at the surface, or gas flares naturally from vents. Spectacular gas seeps in Azerbaijan provide impressive natural fireworks and result from gases vented off from shallow buried reservoirs. Mud volcanoes also provide a clue to hidden hydrocarbons with methane, carbon dioxide and hydrogen sulphide bubbling up in vertical pipes released from liquid mud, similar to a volcanic vent.

© Springer Nature Switzerland AG 2019
M. Searle, *Geology of the Oman Mountains, Eastern Arabia*,
GeoGuide, https://doi.org/10.1007/978-3-030-18453-7_5

Early drilling into shallow oil and gas reservoirs resulted in some devastating natural fires, most notably at Baku in Azerbaijan, in Iraq where IPC first drilled into a major oilfield, and Iran. The 'eternal fires' at Baba Gurgur in the Kirkuk field in northern Iraq have been burning for 400 years. In Oman, the Yibal-14 well blow-out in August 1970 was another classic example of drilling into a highly over-pressurized shallow gas reservoir, resulting in a fire that took over six months to put out.

Oil was discovered in the Middle East over a hundred years ago after surface oil seeps, and even some natural gas flares were found in places, but drilling and production really took off after the Second World War. The first oil discovery in the Middle East was the Masjid Sulaiman field in the southwestern Zagros mountains of Iran, then called Persia, in May 1908. The field was discovered by the Burmah Oil Company which together with Anglo-Persian oil company was later incorporated into British Petroleum (BP). The reservoir was the Oligocene–Early Miocene Asmari limestone and the well struck oil at a depth of 354 m. The large Kirkuk oilfield in Iraqi Kurdistan was discovered in October 1927 by the newly formed Turkish Petroleum Company (TPC), a consortium of Anglo-Persian, Royal Dutch Shell, Companié Française des Pétroles (CFP), the American Near East Development Corporation, and the Gulbenkian Foundation, which as a result of the discovery was renamed the Iraq Petroleum Company (IPC) in 1929. Production in the Kirkuk field began in 1934. This major discovery was soon followed by the discovery of the Jebel Dukhan oilfield onshore in western Qatar extending offshore to Bahrain by the Bahrain Petroleum Company, a subsidiary of Standard Oil of California, in 1932. This was the first major oil discovery in the Late Cretaceous Wasia Group limestone reservoir. The Middle Cretaceous Natih Formation in the Wasia Group was to become the most productive reservoir along the Arabian Gulf in the Oman region. Large oil fields were also discovered in the Euphrates valley of eastern Syria in 1956, in massive Cretaceous limestones, with development beginning in 1969.

The era of major oil discoveries in the Middle East was heralded in 1938 by the discovery of the very large Burgan oilfield in Kuwait in the Albian (Cretaceous) Burgan sandstone reservoir, and in the same year the massive Dammam field in the Upper Jurassic Arab carbonate reservoirs of eastern Saudi Arabia. The Dammam field was discovered and developed by Standard Oil of California that was to become part of Saudi Aramco, the largest oil company in the Arabian Peninsula. Oil and gas was discovered in Qatar in 1940 (Dukhan field) and in Abu Dhabi in 1953 (Bab Murban field). During the 1960s oil and gas drilling technology advanced dramatically and offshore oil-wells were drilled throughout the Arabian Gulf region. In 1960 a massive new discovery offshore northern Qatar was the

Northern Gas field and Al-Shaheen oilfield, with approximately 25 billion barrels of oil.

Abu Dhabi Petroleum Company was formed from a subsidiary of IPC and later became Abu Dhabi National Oil Company (ADNOC). Most of the oil reservoirs in the United Arab Emirates are located within Abu Dhabi territory, although minor amounts have subsequently been discovered in Sharjah (Sajaa field in December 1980), Dubai, Ajman and Umm al Quwain. Ras al Khaimah has three minor fields offshore the west coast of Musandam, and the Emirate of Fujairah is mostly mountainous and lacking oil and gas fields.

In Oman the first discovery of what would later become a commercial oilfield was the Marmul-1 well in south Oman in 1956, drilled by Dhofar Cities Service Petroleum Corporation. This discovery was, however not commercial at the time, and several partners dropped out of the exploration phase leaving only Shell and Partex (Gulbenkian Foundation). Exploration shifted from the Dhofar region northwards towards the foothills of the northern Oman Mountains. In January 1956 Fahud-1 well was spudded on the most promising of a series of very large-scale anticlines (Fig. 5.1). Fahud-1 missed the pay-zone by 200 m, and was dubbed the 'unluckiest well in history' by Julian Paxton (Fig. 5.2). It was not until the second well was drilled at Fahud-2 in 1964 that the major Fahud oilfield in the Natih Formation (Wasia Group) was discovered (Fig. 5.3). The original well was abandoned in 2005 after producing more than five million barrels of oil from this reservoir. The Fahud-2 well-head is on display outside PDO's Oil and Gas Exhibition Center in Mina al Fahal (Fig. 5.4). During the period 1963 to 1967 a series of discoveries were made on large anticlinal structures in Yibal and Natih in 1963, and at Fahud in 1964 (all in Natih Formation reservoirs), and later at Al Huwaisah (Shuaiba Formation reservoir) in 1969 (Hughes-Clarke 1990).

Exploration focus shifted to Central Oman, South Oman and in particular the Ghaba salt basin around the emergent salt domes west of the Haushi-Huqf 'high'. In 1972–3 major oil fields at Qarn Alam, Saih Nihayda, Amal and Saih Rawl were discovered. The oil pipeline to Yibal, Fahud, Natih and Al Huwaisah was extended further south to tap into these central Oman fields. In 1974 a major restructuring of PDO came when the Government of Oman took a 25% stake in PDO, soon to rise to 60%. Shell retained the major outside stakeholder (34%) with CFP, now Total (4%) and Partex (2%) minor players. Shell was still the main operators and carried out all the exploration and production work. One early setback was the oil blow-out of the well Yibal-14 in 1978 when a shallow reservoir ignited and blew out the rig, setting the field ablaze (Figs. 5.5 and 5.6). The fire took 4 months to put out by a specialised blow-out crew under Red Adair and Boots Hansen from Houston.

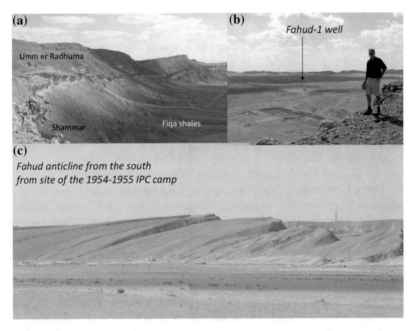

Fig. 5.1 **a** The double-plunging anticline of Jebel Fahud with Paleocene Umm er Radhuma limestones overlying Fiqa shales in the core of the anticline; **b** site of the original Fahud-1 well; **c** Jebel Fahud anticline viewed from the original IPC camp

In 1979 the large Rima field was discovered, and in 1980 the Nimr field followed with the huge network of wells at Marmul in the south brought on line. All these fields produced from Permian glacial deposits (Al-Khlata Formation). By 1997 Oman's oil production peaked at 846,000 barrels per day. During the 2000s a major new initiative was launched to increase oil production in maturing fields. These included a steam injection project at Qarn Alam and a miscible gas injection project at Harweel. More than 20% of oil production now comes from these Enhanced Oil recovery projects (Heward 2006). In south Oman oil is produced from Precambrian rocks, the oldest such reservoirs in the World.

Transporting the oil to the coast where it can be pumped into tankers has always proved a major problem. After the discovery of large oil fields in Fahud, Natih, Yibal and Al Huwaisah, a pipeline was constructed from the coast at Mina al Fahal, near Muttrah, across the mountains through the Semail Gap. Pumping stations were required along the way to get the oil to flow over the watershed and up to the

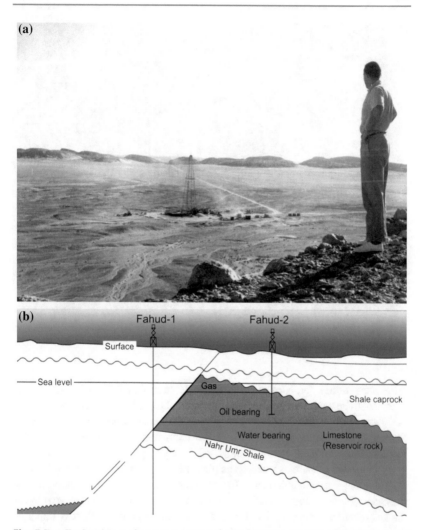

Fig. 5.2 a Exploration geologist Peter Walmsley looking over the Fahud anticline, site of Fahud-1 well; **b** cross section across the Fahud structure showing the location of Fahud-1 which narrowly missed the pay zone, and Fahud-2 well drilled in 1964, which struck the oilfield reservoir

Fig. 5.3 The Fahud-2 well site

large storage tank farms in the hills above Mina al Fahal. The oil pipeline now extends all the way from Marmul and beyond to the north coast. Oil is piped nearly 1000 km north across the desert plains, over the Oman mountains, and along the Semail Gap to Mina al Fahal, where the main office headquarters of PDO is located. From the tank storage facility, oil is fed by gravity to three large single buoy moorings offshore, to which the giant tankers are moored and can load.

Three important requirements for successful oil and gas exploration are the presence of a good **source rock**, a suitable **reservoir** and an effective **seal** to trap the hydrocarbons (Al-sharan 2014; Al-Kindy and Richard 2014). It is important that these components are assembled at the right time; a trap must be present before the oil or gas migrated into it. The relative timing of deformation and expulsion of hydrocarbons from the source are critical. Crude oil and gas are the products of the thermal breakdown of organic matter due to increasing temperature through burial. To accumulate sufficiently rich organic matter in the first place, it must accumulate under anoxic (oxygen-free) conditions. These anoxic environments are usually restricted basins like the present-day Black Sea. Subsidence results in the slow

Fig. 5.4 The original Fahud-2 well-head now displayed outside the PDO Oil and Gas Exhibition center at Ras al Hamra; Bruce Levell for scale

Fig. 5.5 Oil well blow-out at Yibal-14 well, July 1970, photo by Geoff Searle

Fig. 5.6 Fires burning at the collapsed oil rig Yibal-14, July 1970, photo by Geoff Searle

compaction of these organic-rich sediments and over time these sediments begin to mature and form oil. Oil maturation occurs in a specific oil 'window' where the temperature and depth are right for preservation. At depths greater than about four kilometers, or temperatures greater than 120 °C, oil generation will decline, and oil is replaced by gas. Below depths of about 6 km and temperatures greater than 180 °C conversion of the organic matter to carbon is virtually complete and even gas generation will no longer occur. Once liquefied, oil will flow on a grain scale, or via cracks and fractures into reservoirs that can be formed by faulting or folding. The oil then flows and migrates through a porous medium (e.g. sandstone, limestone) or through micro-cracks. For a structure to retain oil it has to have an impermeable cap-rock (e.g. shale) to keep the oil trapped beneath. In a typical Middle East oilfield the reservoir horizon lies above the source and is capped above by a seal. Classic oil and gas structural traps include faults, large domal anticline structures and around salt domes. In some fields however, there is no anticline or structure other than a simple tilt (dip) but the reservoir is discontinuous and sealed above, below and along the sides by a seal, the so-called 'stratigraphic traps'.

Although important oil reservoirs occur throughout the Late Precambrian to Cenozoic time, over 70% of Middle East oil is found in and was formed in the Jurassic–Cretaceous dominantly carbonate sedimentary rocks. In the United Arab Emirates and Oman important stratigraphic source horizons occur in the Neoproterozoic, Lower Silurian, Lower Devonian, Upper Jurassic, Lower and Middle Cretaceous and Palaeogene. Major reservoir horizons are in Infra-Cambrian, Cambrian–Lower Ordovician, Permian, Triassic, Jurassic, Cretaceous and Paleogene. The stable, passive continental margin of Arabia, formed during the Permian as the southern super-continent Gondwana, including Arabia, rifted away from the northern super-continent Laurasia, forming the Tethyan Ocean in between. This passive margin lasted from the Middle Permian, throughout the Mesozoic until the Late Cretaceous time, a period of approximately 150 million years of stable, shallow marine conditions during which approximately three to four kilometers thickness of dominantly carbonate sediments (limestones, dolomites, calcareous shales) full of marine organisms accumulated.

During the Jurassic and Cretaceous periods the shallow Arabian continental margin was over 2000 km wide, rich in organic material and located close to the Equator in warm tropical seas. Oxygen isotope analyses show that the Middle Cretaceous was a time of severe global warming due to a rapid increase in atmospheric carbon dioxide. Warm climate and stable tectonic conditions led to increased nitrogen, phosphorus and carbon contents of the oceans, which in turn led to an explosion in planktonic life, a requirement for organic hydrocarbon formation. The stratigraphic column shows that eastern Arabia and the Gulf region has extensive source rock horizons and extensive seal horizons to trap the hydrocarbons.

Effective seals are marine shales and marls, and evaporates (gypsum, salt, anhydrite). Evaporite horizons form particularly effective seals to trap the oil in lower stratigraphic horizons and occur throughout the Phanerozoic rock record from Arabia when sea-level dropped or when the continental shelf rose above sea-level. In eastern Arabia and the Gulf region the most important evaporite seal horizons are the Infra-Cambrian Hormuz (Ara) salt, the Lower Silurian, Upper Jurassic, several horizons in the Lower and Middle Cretaceous and the Miocene Gachsaran–Fars Formation.

References

Al-Kindy MH, Richard P (2014) The main structural styles of the hydrocarbon reservoirs in Oman. In: Rollinson HR, Searle MP, Abbasi IA, Al-Lazki A, Al-Kindy MH (eds) Tectonic evolution of the Oman Mountains. Geological Society, London, Special Publication vol 392, pp 447–460

Alsharan AS (2014) Petroleum systems in the Middle East. In: Rollinson HR, Searle MP, Abbasi IA, Al-Lazki A, Al-Kindy MH (eds) Tectonic evolution of the Oman Mountains. Geological Society, London, Special Publication vol 392, pp 361–408

Heward A, Al Rawahi Z (2008) Qarn Sahmah salt dome. Geological Society of Oman Field Guide no. 27

Heward A, Vahrenkamp V, Homewood P, van der Berg M, Mazrui M, Follows E (2006) Spudding-in of Fahud-1, 18 January 1956, 50th anniversary field trip to Jebel Fahud. Geological Society of Oman Field Guide No 16

Hughes-Clarke M (1990) Oman's geological heritage. Stacey Interntional Publishers, p 247

Paxton J (unpublished) History of PDO. Middle East Archive Center, St-Anthony's College, Oxford

Part III
Regional GeoPark Sites

Musandam Peninsula and Straits of Hormuz

The Musandam peninsula is the northernmost promontory of the Oman–UAE Mountains and one of the most distinctive geological features of eastern Arabia (Figs. 6.1 and 6.2). The Straits of Hormuz between Kumzar in Omani Musandam and Qishm Island in Iran is only 50 km wide. The Straits of Hormuz separates the Arabian Gulf to the west from the Gulf of Oman (or Sea of Oman) to the east. The 650 km winding coastline has some unique geomorphology with steep cliffs dropping straight into the ocean, large fjords cutting deep into the mountains and numerous sandy bays and small islands. Late Cenozoic eastward tilting of the entire Musandam peninsula has resulted in a spectacular geomorphology with drowned wadi systems along the north and east coasts creating fjords with sea-cliffs hundreds of meter high. The highest peaks of the Musandam, including Jebel Yibir, are just over 2000 m above sea-level and lie only 4–6 km from the east coast. The main wadis drain westward to Ras al Khaimah (Wadi al Bih, Wadi Ghalilah, Wadi Sham) and northwest to Khasab (Wadi Sal al A'la). Two large fjords cut through the northern Musandam mountains, Khawr As Shaam from the Arabian Gulf and Khawr Habalayn–Khawr Najd from the Gulf of Oman side. A narrow isthmus of mountainous land less than 900 m wide separates the two fjords. The fantastic winding coastline of the northern Musandam testifies to the Quaternary and Recent tilting of the entire peninsula to the east and north. The Royal Geographical Society carried out a major geographical research project in the Musandam peninsula during 1971–1972 (Cornelius et al. 1973).

Most of Musandam lies in Oman but the western mountains belong to the Emirate of Ras al Khaimah. The main town is the port of Khasab in the north with the old fishing villages of Bukha on the west coast and Bayah in the southeast, both now developed into larger towns. Very small mountain villages, now largely deserted, are scattered across the Musandam mountains. These were the homes of the mountain Shihu tribesmen who carry a distinct small axe termed a *jinz*. The

© Springer Nature Switzerland AG 2019
M. Searle, *Geology of the Oman Mountains, Eastern Arabia*,
GeoGuide, https://doi.org/10.1007/978-3-030-18453-7_6

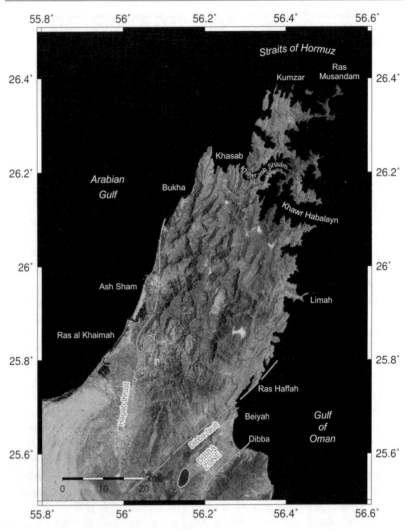

Fig. 6.1 Landsat photograph of the Musandam peninsula, Straits of Hormuz and Dibba zone

Shihu are Arabic speakers, whereas the Kumzaris who live in the far north of Musandam, speak a dialect that is half Arabic and half Persian. Kumzar is the northernmost village in Oman and only has access from the sea, with mountains and fjords blocking any access to the main Musandam region to the south.

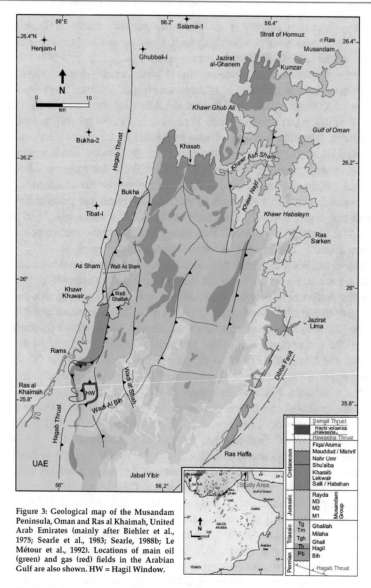

Figure 3: Geological map of the Musandam Peninsula, Oman and Ras al Khaimah, United Arab Emirates (mainly after Biehler et al., 1975; Searle et al., 1983; Searle, 1988b; Le Métour et al., 1992). Locations of main oil (green) and gas (red) fields in the Arabian Gulf are also shown. HW = Hagil Window.

Fig. 6.2 Geological map of the Musandam peninsula

The Musandam mountains are composed almost entirely of Permian and Mesozoic carbonates of the shelf sequence, similar rocks as those exposed in the Jebel Akhdar and Saih Hatat ranges in the central and eastern mountains (Fig. 6.3). A regional unconformity underlies the Upper Permian–Lower Triassic Bih Formation dolomites, and the Middle and Upper Triassic Hagil and Ghail Formations. These massive well-bedded dolomites are equivalent to the Permian Saiq Formation and Triassic Mahil Formation in the central and eastern Oman Mountains. The Bih Formation dolomites formed in a shallow marine tidal flat environment similar to that seen in coastal Abu Dhabi today, and is equivalent to the highly productive hydrocarbon-bearing Khuff Formation in interior Oman. The Triassic Milaha Formation grades from open shelf bioclastic *Megalodon*-rich limestones in northern Musandam to outer shelf skeletal and peloidal lime sand shoals and coral-algal reefs along Batha Mahani, the southern margin of Musandam. The Upper Triassic Jebel Wasa facies is a distinctive reef carbonate containing abundant corals, sponges, encrusting algae and foraminifera.

Musandam is a giant thrust-fold culmination of the Permian to Cretaceous shelf carbonates and the thrust along the base (Hagab thrust) exposed in the Hagil Window in Ras al Khaimah extends deep down into the pre-Permian basement (Fig. 6.4; Searle 1988; Searle et al. 2014; Ali et al. 2017). Spectacular folds in the uplifted Mesozoic shelf carbonate sequence are exposed along the fjordland cliffs (Fig. 6.5). The northern part of the Musandam Peninsula plunges to the north beneath the Straits of Hormuz to the orogenic syntaxis where the Oman–UAE mountains strike into the eastern extension of the Zagros Mountains in Iran.

The Musandam Peninsula is unique in that it records the first effects of a continent–continent collision, as Arabia started to collide and indent into Central Iran during the Miocene (Biehler et al. 1975; Ricateau and Riche 1980; Searle et al. 1983; Searle 1988). Further west this collision has resulted in the Zagros fold-thrust belt in southern Iran. The Musandam shelf carbonates have been uplifted along the Hagab thrust—an Oligocene–Miocene west-vergent thrust fault that effectively cuts up-section from basement to Oligocene beds offshore.

Not only is the Musandam a unique geological and archaeological (Cornelius et al. 1973; Costa 1991) site, but also its coastline is very important for marine life. The inlets around Khawr Ghubb Ali and Khawr As Shaam on the western side and Khawr Habalyn on the eastern side of Musandam have pristine coral reefs, amongst the most northerly reefs in the Indian and Pacific Oceans. The headlands off the east coast of Musandam are well known for schooling sharks of at least 6 different species including the World's largest, the Whale Shark (*Rhinodon typus*) and breaching Manta rays (*Manta birostris*). Rare species of sailfish (*Istiophorus* sp.), guitarfish (*Actroteriobatus* sp.) and sawfish (*Anoxypristis* sp.) are also known

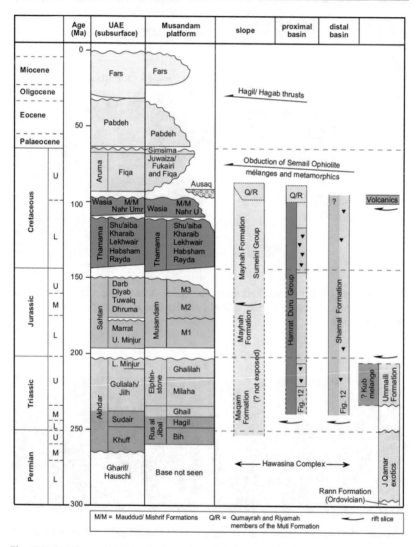

Fig. 6.3 Permian and Mesozoic time chart showing stratigraphy of the UAE foreland, Musandam shelf carbonate platform and allochthonous Tethyan ocean basin (Dibba zone) sedimentary rocks

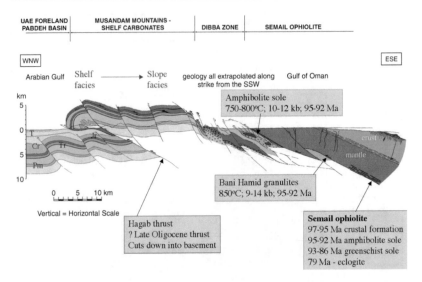

Fig. 6.4 Cross-section of the Musandam mountains, Dibba zone and UAE ophiolite, showing timing constraints on the Semail ophiolite and metamorphic sole rocks from radiometric dating

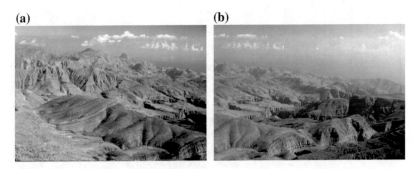

Fig. 6.5 View north and east across the mountains of Musandam from the summit of Jebel Hagab

to be present and schools of dolphins (*Delphinus delphis*) are common sights in the waters off Musandam. Many of the offshore rocky islets have world-class dive sites with cliffs, caves, and overhangs carpeted with a rich and colourful variety of corals and swarms of schooling fish (Salm and Baldwin 1991). Onshore, the

Musandam mountains are known to contain Arabian gazelle (*Gazella gazella*) as well as the rare caracal (*Caracal* sp.) and wolf (*Canis lupus*). It is still possible, although unlikely, that Arabian Tahr (*Hemitragus jayakari*) and Arabian leopards (*Panthera pardus*) inhabit the high more remote mountains. The last leopard sighting was in 1995 when a full-grown male leopard was shot by a herdsman east of Khasab. Musandam is a unique and special place both for its geology and its marine wildlife and it has been proposed that the entire north and east coast should become a World Heritage Site and National GeoPark.

6.1   Hagil Window, Ras al Khaimah

The Hagil Window east of Ras al Khaimah is an extremely important site as this is where the Hagil Thrust is exposed around a large bowl surrounded by high mountains (Fig. 6.6). Wadi Hagil cuts a narrow gorge into the mountains and the

Fig. 6.6 Panorama across the Hagab thrust, Hagil Window, Ras al Khaimah, UAE

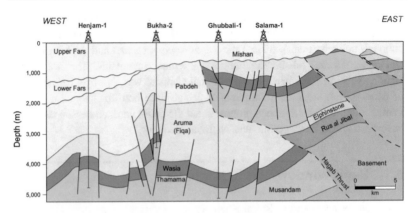

Fig. 6.7 Interpretation of seismic sections offshore Ras al Khaimah, linked to log sections of the Henjam, Bukha, and Ghubbali wells in the Arabian Gulf

entire Musandam shelf carbonate sequence has been folded in a very large-scale recumbent anticline above the Hagab Thrust. The Hagab thrust is major west-vergent thrust fault that repeats the entire pre-Permian basement and Permian–Cretaceous shelf carbonates and cuts up-section as far as the Lower Miocene in the sub-surface (Ali et al. 2017). The Musandam culmination and Hagab thrust phase of compression and uplift from Late Oligocene to Early Miocene reflects the first phase of continent–continent collision as Arabia and Iran converge. The Arabian Gulf to the west is underlain by Arabian continental crust, whereas the Gulf of Oman to the east offshore Oman is underlain by Cretaceous ocean crust that is being subducted northwards beneath the Makran accretionary prism.

The Hagab thrust can also be traced offshore Ras al Khaimah on seismic profiles that can be correlated with oil wells at Henjam, Buhka, west of the Hagab thrust, and Ghubbali-1 and Salama-1 wells that drilled into the hangingwall of the Hagab thrust (Fig. 6.7). The Hagab thrust is well exposed all around the Hagil window with footwall rocks including red cherts, alkaline volcanics and rare alkaline ultramafic rocks of the Haybi complex. A cross-section and restored section across the Hagab thrust and Musandam culmination are shown in Figs. 6.8 and 6.9. A recent well, Hagil-1, was drilled south of As Sham village in Ras al Khaimah emirate which penetrated the Hagab thrust at depth (Ali et al. 2017).

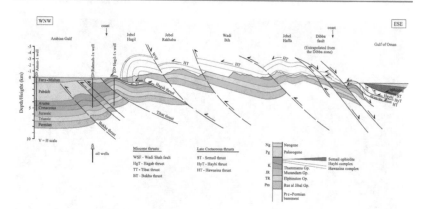

Fig. 6.8 Cross-section of the northernmost Oman Mountains across the Musandam peninsula

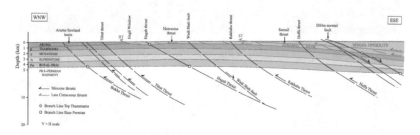

Fig. 6.9 Restored section across Musandam peninsula

6.2   Wadi Bih, Jebel Jais

Wadi Bih is one of largest wadis that cuts into the Musandam mountains from Ras al Khaimah and a new road now links Wadi Bih with the main Khasab–Dibba road along the central part of the peninsula. Wadi Bih provides an excellent transect across the Musandam shelf carbonates and crosses the border into Oman. The watershed peaks along the Musandam peninsula run north–south to the east of upper Wadi Bih. Spectacular cliffs rise from here up to 1900–2000 m high peaks along the watershed, including the highest peaks of Musandam, Jebel Hagab (Figs. 6.10 and 6.11) and Jebel Qihwi (Fig. 6.12). Wadi Shah is a smaller wadi branching off to the north from lower Wadi Bih. Excellent exposures

Fig. 6.10   Jebel Hagab, Musandam mountains

Fig. 6.11   Jebel Hagab, view from summit of Jebel Qihwi

Fig. 6.12 Jebel Qihwi summit, Musandam; Dibba bay and the ophiolite mountains of UAE in the distance

of west-vergent folds and thrusts can be seen along this road that rises to the summit ridge of Jebel Jaiz (Fig. 6.13). Superb views westward down to Wadi Ghalilah (Wadi Litibah) and the west coast of UAE can be seen from the summit ridge. On the Ras al Khaimah side, a trekking route, called the 'Stairway to Heaven', originally made by Shihu tribesmen, ascends the very steep to vertical cliffs from the wadi floor up to Jebel Bil'aysh and Jebel Jaiz. The climb takes between 8 and 10 hours and requires a detailed knowledge of the route, plenty of water, and a head for heights. Excellent rock climbing, trekking and scrambling routes exist all over the Musandam but stricter controls along the Oman–UAE border have now made many of these routes difficult to access.

6.3 Bukha–Harf

The coastal road from Ras al Khaimah to Khasab crosses the border into Oman just north of Shams village. Here the mountains run straight into the sea and impressive steep sea-cliffs rise straight out of the ocean. The new road has been built along the coast and passes some beautiful sandy coves. The large settlement of Bukha has now grown from a sleepy fishing village cut off from the rest of the country by

Fig. 6.13 Frontal fold of Musandam, view northwest to As Sham and the Arabian Gulf

steep cliffs and high mountains to a town with the construction of the coastal road. The town is dominated by the spectacular Bukha fort, built in the 17th century and restored in 1990. Another smaller fort, Al Qala'a fort lies above the town on the eastern side. The cliffs to the north of Bukha show excellent exposures of various splays of the Hagab thrust cutting up through the Jurassic (Musandam Group) and Cretaceous (Salil/Habshan, Lekhwair, Kharaib, Shu'aiba, Nahr Umr and Mauddud Formations) shelf carbonates (Fig. 6.14a). The main Hagab thrust runs north out to sea in between the Tibat-1 and Bukha-2 wells to the west (footwall) and the Ghubbali-1 well to the east (hangingwall; Fig. 6.14b).

From Bukha the main road contours along the coast in-between the sea and the cliffs for 15 km before crossing the northwestern promontory and Harf plateau to Khawr Qida, a small fishing hamlet. From Harf village there are great views north across the Straits of Hormuz to Qishm Island in Iran, and some nice treks to access rocky beaches by steep rocky gullies. The road continues along the coastline to Khasab, the main town of Musandam at the head of a large shallow inlet, or khawr. From the port area of Khasab it is easy to hire dhows to go to Khawr As Shaam, and further north along the amazing west coast of Musandam to Jazirat Ghanem and Kumzar, the northernmost village in Oman.

Fig. 6.14 Musandam photos, **a** panorama across the high peaks of Musandam, **b** imbricate thrusts above the Hagab thrust, Bukha, Ras al Khaimah, **c** folding in Triassic–Jurassic shelf carbonates, Khawr As Sham, Musandam

6.4 Khawr As Shaam

Khawr As Shaam is a spectacular and impressive fjord cutting approximately 20 km eastward from Khasab into the Musandam mountains. Steep sided cliffs drop straight into the sea and only two tiny fishing settlements of a few houses occupy very small beach areas along the north shore. The sea is rich in life with large pods of dolphins very commonly seen as well as manta rays, reef sharks and the occasional whale. Coral reefs occur in patches along the khawr and attract reef fishes of every colour and size. At the eastern end of the khawr a small island, Telegraph island, is surrounded by coral reefs. Telegraph island was the site of a cable station used by the British from 1864 to 1869, and the ruins of some of the buildings can be see today. It is possible to hire a dhow or small boat from Khasab for a days' trip around Khawr As Shaam, including swimming and snorkelling stops. There is a beautiful remote white sand beach on the south shore of Khawr As

Shaam that provides a magnificent camping spot a few kilometers east of Khasab. East of Telegraph island the far end of Khawr As Shaam is an isthmus only 900 m across that separates Khawr As Shaam from Khawr Habalyn and the Gulf of Oman. Geologically the shelf carbonates show steep to west-vergent folds and two thrust faults that place Triassic brown limestones of the Ghalilah Formation over Jurassic limestones of the Musandam Group (Fig. 6.14c).

6.5 Kumzar

The northernmost peninsula of Musandam is an amazing chaos of contours around the drowned wadi systems that drain north into the Straits of Hormuz. As the Musandam peninsula tilted north and eastwards during the Pleistocene and Quaternary the wadis became drowned. The mountains drop straight into the sea and there is only one beach where the settlement of Kumzar has been built (Fig. 6.15). Kumzar is completely cut off from the rest of Oman by mountains and a tortuous coastline. The only way in or out of Kumzar is by sea. The seas are always

Fig. 6.15 Kumzar village, north coast of Musandam

extremely rough and dhows from Khasab often have to turn back due to mountainous waves and strong currents. Jazirat al Ghanem ('goat island') is a large island off the west coast of the Kumzar peninsula. This was the site of a British telegraph station during the last century and a few ruined buildings are still present there. Between Kumzar and Khasab a large fjord, Khawr Ghub Ali cuts east into the Lower Cretaceous and Upper Jurassic limestones with steep or overhanging sea-cliffs all around. There is no easy landing place for dhows anywhere in Khawr Ghub Ali or along the coast to Kumzar.

The mountains around Kumzar are composed of Jurassic Musandam Group limestones to the east and the Cretaceous sequence to the west. The giant west-verging fold of Musandam plunges gently to the north beneath the Straits of Hormuz towards a syntaxis point in the middle of the Straits, south of Qishm island. The Upper Triassic Ghalilah Formation and the Jurassic Musandam Group shelf carbonates are beautifully exposed in the mountains around upper Wadi Bih (Fig. 6.16). The entire Arabian Gulf drains out to the Gulf of Oman through the Straits of Hormuz, between Oman and Iran which is only about 50 km wide at this point. Between 20 and 30 oil tankers a day pass through the Straits of Hormuz

Fig. 6.16 Musandam photos, **a** dip slope in Triassic Milaha and Ghalilah Formations, Wadi Bih, **b** full section through the Jurassic Musandam Group limestones, Wadi Bih, **c** Jurassic Musandam Group limestones, Jebel Hagab, **d** Jurassic Musandam group limestones around Kumzar village, far north coast of Musandam

from oil ports in Iraq, Kuwait, Saudi Arabia, Qatar and UAE, so Musandam and Kumzar has become a strategically very important place.

6.6   Sal Al A'la, Khawr Najd, Khawr Habalyn

East of Khasab, a large wide wadi runs eastward cutting thorough all the mountains except for the last 2 km to the Gulf of Oman. It is possible to climb a ridge approximately 450 m high and trek over these mountains from the eastern end of Wadi Sal Al A'la to gain spectacular views along the east coast to the Gulf of Oman. There are abandoned terraced fields from old Shihu settlements in some of these valleys. The coastline is a series of rocky headlands and rocky beaches from where it is possible to swim and snorkel along the cliffs. There is a small forest of acacia trees at Al Khalidiya in Wadi Sal Al A'la which makes a pleasant camping spot. Half way between Khasab and Al Khalidiya, a road branches north and climbs over a ridge to descend to another spectacular inlet, Khawr Najd via a series of zig-zag bends. Khawr Najd is an impressive fjord with steep sea-cliffs lining both sides, eventually connecting north to Khawr Habalyn and the Gulf of Oman. There are no villages or settlements anywhere along Khawr Najd or Khawr Habalyn so it is only possible to see this remote and incredibly scenic fjord by boat.

6.7   Limah and East Coast

The east coast of Musandam is one of the most spectacular coastlines anywhere in the World. Although the Musandam peninsula is presently tilting toward the north and east as evidenced from the spectacular drowned coastline, the geology shows the opposite, a westward tilt with the oldest rocks exposed along the east coast. Upper Permian limestones and dolomites of the Bih Formation and Lower Triassic Hagil and Ghail Formations are exposed along the southeast coast of Musandam from Ras Haffa to Jazirat Lima. Steep sea-cliffs drop into the Gulf of Oman from the watershed ridge at 1700–2000 m altitude with rocky headlands, tiny coves and remote fjords. The headlands jutting east into the Gulf of Oman are unique diving sites with many species of sharks, rays and whales breeding in this area. Upwelling during the winter months provide nutrient-rich waters for the larger fish and the whole seas here form a rich fishing ground. Strong ocean currents sweep along the east coast of the Musandam peninsula. Limah is a tiny settlement, no longer inhabited, located on the largest headland. It is possible to land only at a few localities along this coastline, and only by swimming ashore from the dhow.

6.8 Ras Haffah and Zighy Bay

The southeastern part of Musandam is the large Dibba fault, a steep east-southeast dipping normal fault that bounds the southern part of the Musandam shelf carbonates to the north from the Dibba zone allochthonous Hawasina and Haybi complex rocks to the southeast. Numerous quarries have now cut out a huge amount of limestone for building projects in the UAE. Rocks quarried from here form the foundations for the Palm Islands, offshore Dubai, and Jumairah and reclaimed land offshore Abu Dhabi. The Dibba fault splays into several strands north of Beiyah and Dibba. One branch runs through a beautiful beach, Zighy Bay, now the site of a 5-star resort and a small local fishing village. The fault runs right up along the coast offshore Ras Haffah. Another branch runs through the Omani village of Beiyah, north of Dibba. The Dibba faults runs from here southwest along Batha Mahani dividing the complex folded and thrust rocks of the Dibba zone from the massive shelf carbonates of Musandam to the northwest. This fault may be the surface expression of the continent-ocean crust boundary in the Gulf of Oman. East of the Dibba fault the Semail ophiolite was obducted from the Tethyan oceanic realm, whereas west of the Dibba fault is purely continental crust. The geology of the Gulf of Oman as revealed by seismic studies shows a thick Cenozoic sedimentary package overlying some sort of basaltic oceanic substrate, interpreted as Gulf of Oman Cretaceous oceanic crust. The Cenozoic sediments have been folded by young compressional forces thought to be the affects of the collision of the Arabian plate with the Iranian plate. In the Gulf of Oman the present day plate margin is thought to be a shallow north-dipping subduction zone south of the Baluchistan coast.

References

Ali MY, Aidarbayev S, Searle MP, Watts AB (2017) Subsidence history and seismic stratigraphy of the Western Musandam Peninsula, Oman–United Arab Emirates. Tectonics 37. https://doi.org/10.1002/2017TC004777

Baldwin R, Salm R (1991) Snorkelling and diving in Oman. Motivate Publishing

Biehler J, Chevalier C, Ricateau R (1975) Carte Geologique de la Peninsula de Musandam. BRGM, Orleans, France

Costa PM (1991) Musandam—architecture and material culture of a little known region of Oman. Immel Publishing, London, 249 p

Cornelius PFS, Falcon NL, South D, Vita-Finzi C (1973) The Musandam expedition 1971–2, scientific results. Geogr J 139:400–425

Ricateau R, Riche PH (1980) Geology of the Musandam Peninsula (Sultanate of Oman) and its surroundings. J Pet Geol 3:139–152

Searle MP (1988) Structure of the Musandam culmination (Sultanate of Oman and United Arab Emirates), and Straits of Hormuz syntaxis. J Geol Soc Lond 145:43–53

Searle MP, James NP, Calon TJ, Smewing JD (1983) Sedimentological and structural evolution of Arabian continental margin in the Musandam Mountains and Dibba zone, United Arab Emirates. Geol Soc Am Bull 94:1381–1400

Searle MP, Cherry AG, Ali MY, Cooper DJW (2014) Tectonics of the Musandam Peninsula and northern Oman Mountains: from ophiolite obduction to continental collision. GeoArabia 19:135–174

United Arab Emirates (UAE)

7

The United Arab Emirates has some stunning scenery, not only in the northern part of the Oman–UAE mountains but also in the desert foreland. Giant red sand dunes stretch off towards the Empty Quarter and Saudi Arabia from the Liwa oasis, south of Abu Dhabi. The western coastline is flat and prone to tidal inundation, with stretches of white sand beaches and sabkhas making up most of the coast. Dubai, Sharjah, Umm al Quwain and Ras al Khaimah are all cities build around creeks or *khawrs*. These creeks were centers of the pearl diving industry and became important dhow trading ports with trade routes established all over the Gulf and across to Iran, Baluchistan, and India, and even further afield to Kenya, the islands of Pemba and Zanzibar and Tanganyika. Ras al Khaimah was an important port for the Trucial coast pirates more than a century ago, and Dubai became a center for gold smuggling. The Emirati cities have grown from poverty-stricken fishing villages with palm frond *barasti* huts, to 21st century cities with the World's largest skyscrapers in just forty years.

Most of this coastline has been highly developed and at least four offshore 'Palm Islands' have been constructed from reclaimed land offshore Dubai, including the 'World'. These artificial islands have been built from rocks quarried from southern Musandam. Several islands offshore Abu Dhabi are emergent salt domes. One of them, Sir Bani Yas is now a nature reserve and has large herds of African and Arabian species. Offshore UAE, whale sharks, manta rays and dugongs jostle for space between the oilrigs and supertankers. Fujairah is the only Emirate entirely in the mountains and along the east coast, although many Emirates have two or three separate pieces of territory in the mountains. The UAE part of the mountains includes Fujairah and some enclaves of Sharjah and Ras al Khaimah occupying the northern mountains south of Omani Musandam and north of Wadi Hatta. The main geological sites of interest are described below. Figure 7.1 is a Landsat satellite photo of the UAE mountains and Fig. 7.2 is an aerial photo looking across the UAE mountains towards the Musandam mountains in the north.

© Springer Nature Switzerland AG 2019
M. Searle, *Geology of the Oman Mountains, Eastern Arabia*,
GeoGuide, https://doi.org/10.1007/978-3-030-18453-7_7

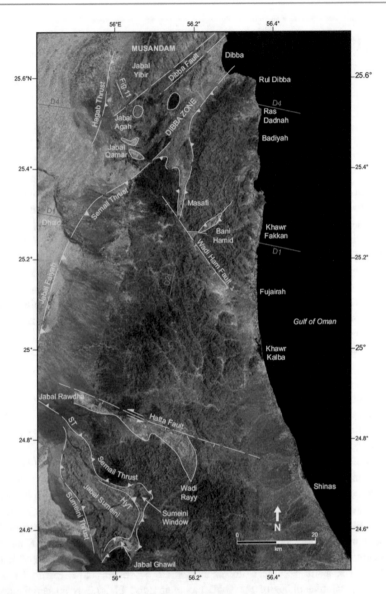

Fig. 7.1 Landsat photo of the northern part of the Semail ophiolite in Oman and UAE. The Dibba zone separates the Musandam shelf carbonates in the north from the ophiolite. D1–D4 shows the lines of section for the geophysical profiles

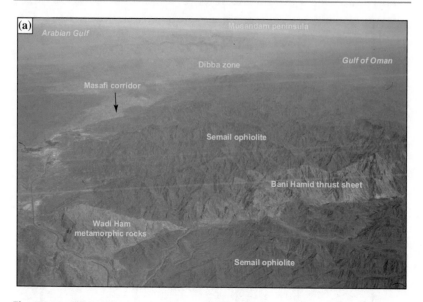

Fig. 7.2 Aerial photograph looking north, showing the northern part of the Semail ophiolite in UAE, the Masafi corridor and Wadi Ham–Bani Hamid metamorphic thrust sheets

7.1   Dibba Zone

The Dibba zone is a structurally complex zone separating the Musandam shelf carbonates to the northwest from the Semail ophiolite to the south. It consists of a series of WSW-verging thrust sheets of more proximal Hamrat Duru Group (Dibba and Dhera Formations) and distal Hawasina complex (Shamal cherts) thrust sheets overlain by the more distal Haybi complex. The Haybi complex thrust sheet includes the Jebel Qamar exotic limestones, the Kub mélange and imbricated thrust sheets of greenschist and amphibolite facies metamorphic sole (Fig. 7.3). Since the 1980s, many of the key geological localities have been destroyed through unregulated construction development and new roads. The entire thrust stack exposed in the Dibba zone has been folded on a large scale, such that the structurally deepest rocks are exposed in inliers such as Jebel Agah. A cross-section of the Dibba zone is shown in Fig. 7.4 and a restoration of these rocks on a profile from the passive continental margin across the proximal slope to the distal basin is shown in Fig. 7.5.

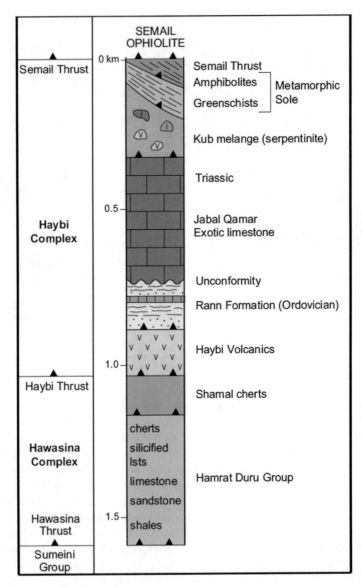

Fig. 7.3 Composite structural–stratigraphic profile through the Haybi complex and distal Hawasina complex rocks of the Dibba zone, the likely protolith of the Bani Hamid granulites

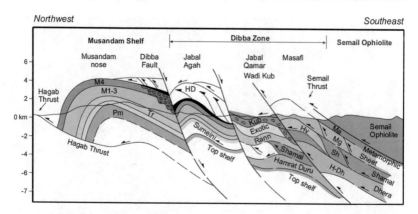

Fig. 7.4 Simplified geological cross-section across the Dibba zone showing major thrust sheets underlying the Semail ophiolite

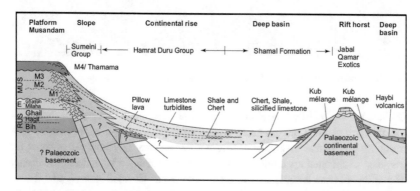

Fig. 7.5 Restoration of the thrust sheets in the Musandam–Dibba zone, UAE showing a palinspastic reconstruction of the shelf-slope-basin profile

Extensive outcrops of alkali basalts, ankaramites and trachytes have been mapped across the Dibba zone, similar to Haybi complex outcrops further south in Oman (Searle et al. 1983; Maury et al. 2003; Chauvet et al. 2009). Unusual rocks include carbonatites, alkali peridotite (jacupirangite, wehrlite) and pyroxenite sills have also been found (Searle et al. 1983), indicative of the deep levels of a within-plate, off-axis ocean island. The Haybi complex volcanic rocks are dominantly Middle Permian to Late Triassic age (Fig. 7.6), but some pillow lavas in the

Fig. 7.6 Triassic pillow lavas of the Haybi complex, Dibba zone, UAE

Dibba zone are Middle Jurassic and even Late Cretaceous. Because of their wide regional extent, they are probably related to passive margin rifting, rather than related to any plume magmatism. It is not known precisely how far offshore the extended continental crust underlay the Tethyan oceanic rocks, but the occurrence of mildly alkaline basalts in the distal Hawasina (Shamal cherts) and Haybi complexes suggest that these rocks were underlain by oceanic alkali basalt substrate. The structurally highest thrust slices are composed almost exclusively of thin-bedded radiolarian cherts that were deposited in the distal farthest parts of the preserved Tethyan ocean (Fig. 7.7).

The Dibba zone has undergone extensive developments in the last twenty years, with major road networks, new towns and large quarries. Huge quarries along the northern side of Batha Mahani have provided rocks for offshore island developments in Dubai, and some other quarries have targeted decorative marbles from the Oman Exotics. Many key geological localities described in some older papers, have been destroyed.

Fig. 7.7 Folded radiolarian cherts in the Dibba zone, UAE

7.2 Jebel Qamar Exotic

The Triassic Jebel Qamar south exotic (Fig. 7.8) is a large limestone block cropping out in the middle of the Dibba zone underlain by the continental rocks of the
Ordovician Rann Formation (Fortey et al. 2011). This is the only exotic limestone
in the Oman–UAE mountains to be underlain by a rifted piece of continental
basement, and therefore unlike all the other exotic blocks in the UAE–Oman
mountains. All other exotic limestones in the Haybi complex are underlain by
alkali basalts interpreted as Triassic volcanic substrate to ocean island guyots or
seamounts (Searle and Graham 1982; Pillevuit et al. 1997). Most of the Rann
Formation outcrops have now been quarried away, but a detailed biostratigraphic
study by Fortey et al. (2011) shows that the faunas are dominantly Middle and Late
Ordovician. Jebel Qamar is interpreted as a thin slice of Arabian continental

Fig. 7.8 Old photo (1978) looking across the Dibba zone at folded limestones of Jebel Qamar and Hawasina and Haybi complex rocks underlying the ophiolite

Fig. 7.9 Old photo (1978) showing folded limestones of Jebel Agah and the Jebel Qamar exotic limestone in the Dibba zone, UAE, view towards south from Musandam

basement that rifted away from the margin during Permian–Triassic time and ended up in the distal part of the Tethys ocean with a Triassic carbonate bank on top. The shelf slope carbonates (Sumeini Group) were also folded during the Late Cretaceous along with all overlying thrust sheets of distal Hawasina rocks (Fig. 7.9).

7.3 Masafi Metamorphic Sole

The metamorphic sole of the Semail ophiolite is a narrow sequence of metamorphic rocks showing an inverted thermal gradient welded onto the base of the mantle sequence peridotites during early subduction and obduction. Along the base of the sole, greenschist facies rocks include mainly metamorphosed cherts and minor meta-volcanic rocks, structurally overlain by epidote amphibolites, and garnet + clinopyroxene-bearing amphibolites above. Along the upper part of the amphibolites, tiny enclaves of garnet + diopside granulite facies rocks are enclosed by amphibolite. At the structurally highest levels of the amphibolites, small wispy pods of partial melts have a tonalitic composition and are related to the highest temperatures of metamorphism around 900 °C. A sharp contact with the peridotites (harzburgites, dunites, lherzolites) above marks the position of the Semail thrust. Metamorphism has also affected the basal peridotites with crystallisation of hornblende in the harzburgites and lherzolites and a banded ultramafic unit showing a high degree of shearing and serpentinisation (Ambrose et al. 2018). The metamorphic sole shows high strain mylonitisation, recumbent folding and ductile shearing throughout, but does maintain a structural coherence such that all sections where the sole remains intact are similar. In places the sole has been imbricated during emplacement such that the sequence is tectonically repeated. In other places the metamorphic rocks form boulders enclosed by serpentinised peridotite in a tectonic mélange.

The best section through the metamorphic sole is a small wadi section immediately north of Masafi town. This is one of the most complete sections of sub-ophiolite metamorphic sole in the Oman–UAE mountains, if not the World. The lower part is marked by an upward increase in strain and tight to recumbent folding in green, red and white cherts. Strong shearing fabrics show anastomosing high strain ductile shear zones. One horizon of unusual kyanite-bearing quartzites occurs, where Gnos (1998) obtained temperatures and pressures of about 800 °C and 11 kbar from these rocks. Overlying these quartz-rich mylonites are amphibolites composed of hornblende + plagioclase ± epidote. Garnet and clinopyroxene appear above these rocks and extend up to the Semail thrust. Cowan et al. (2014) and Ambrose et al. (2018) obtained pressure–temperature conditions of

770–900 °C and 11–15 kbar for the garnet-clinopyroxene amphibolites and granulites. The depth of burial (40–50 km) recorded by these sole rocks is far greater than can be accounted for by the thickness of the ophiolite (~ 15–20 km maximum). The high strain and high degree of metamorphism restricted to the basal contact of the ophiolite suggest that these rocks were formed in a subduction zone environment beneath the Semail ophiolite (Searle and Malpas 1980, 1982; Searle and Cox 1999, 2002). U–Pb zircon dating shows that the ophiolite and the sole amphibolites formed at precisely the same time (96.5–95.5 Ma; Rioux et al. 2016) confirming the origin of the ophiolite must have been above an active subduction zone. At Masafi it is possible to traverse a complete and wholly exposed section across the metamorphic sole and up into the peridotites of the mantle sequence. Another good profile through an intact amphibolite-granulite sequence metamorphic sole occurs at Khubakhib, north of Masafi, where ~ 70 m of garnet amphibolite remains attached to the base of the ophiolite (Gnos 1998; Soret et al. 2017).

7.4   Bani Hamid (Madhah) Granulites

The small Omani enclave of Madhah is surrounded by UAE territory but has one village in the middle, Nahwa belonging to UAE. The Bani Hamid thrust slice of high-temperature granulite facies metamorphic rocks emplaced into the peridotites of the northern Oman–UAE ophiolite outcrops in the western part of Madhah, and continues west into UAE towards Wadi Ham (Fig. 7.10). The Bani Hamid granulites are unique in the Oman–UAE mountains and are very different from the sub-ophiolite metamorphic sole rocks anywhere else. They are over 1000 m thick and all within the high-temperature granulite or amphibolite facies, but without the characteristic inverted metamorphic field gradient of other sole outcrops (Fig. 7.10; Searle et al. 2014). Rocks in the Bani Hamid thrust sheet contain some unique mineral assemblages in Oman, including:

(a) Orthopyroxene (enstatite) + clinopyroxene (diopside) quartzites (± hornblende, cordierite, sapphirine),
(b) Marbles and calc-silicates containing diopside + andradite garnet + wollastonite + scapolite,
(c) Amphibolites containing hornblende + plagioclase ± clinopyroxene ± biotite, with localised partial melting,
(d) Hornblende pegmatites.

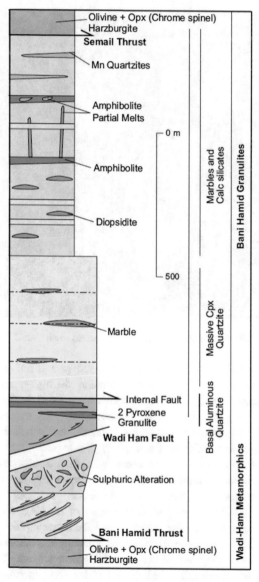

Fig. 7.10 Profile through granulite and amphibolite facies rocks of the Bani Hamid thrust sheet

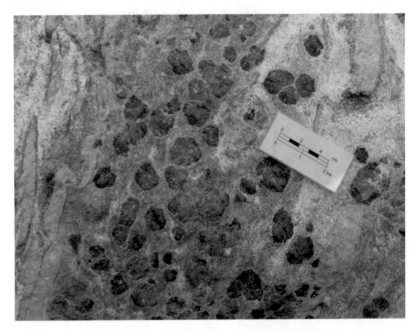

Fig. 7.11 Andradite garnets in scapolite-bearing marbles, Bani Hamid thrust sheet

The marbles and calc-silicates show large red garnets and green pyroxenes floating in a ductile matrix of carbonate including the high-grade minerals wollastonite and scapolite (Fig. 7.11). Intense high-temperature ductile flow folding is apparent (Fig. 7.12). The quartzites show extremely high temperature minerals including both orthopyroxene and clinopyroxene, and sapphirine, an unusual aluminium-magnesium silicate (Gnos 1998). These rocks were derived from metamorphosed distal cherts and carbonates with alkali basalts, so are most likely to be metamorphosed components of the Haybi complex (Searle et al. 2014, 2015). Pressure–temperature conditions of formation are $850 \pm 60\ °C$ and 6.3 ± 0.5 kbar. U–Pb zircon and titanite ages of ca. 94.5–89.8 Ma, slightly younger than the ages from the ophiolite, and the metamorphic sole (Fig. 7.13; Rioux et al. 2016). The entire Bani Hamid granulite-amphibolite sequence has been affected by folding on all scales and high-temperature ductile shearing. Large-scale tight west-verging folding that restores to a minimum shortening length of 130 km with an additional minimum 30 km offset along the Bani Hamid thrust (Fig. 7.14). This is an out-of-sequence thrust that cuts through the crustal stack up into the mantle sequence of the ophiolite. The sequence of events showing

Fig. 7.12 Extreme ductile folding in Bani Hamid granulite-facies marbles

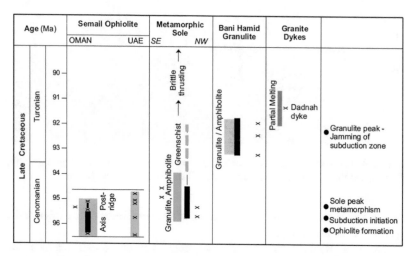

Fig. 7.13 Late Cretaceous time chart showing age ranges of the Semail ophiolite, Metamorphic sole and Bani Hamid granulite thrust sheets

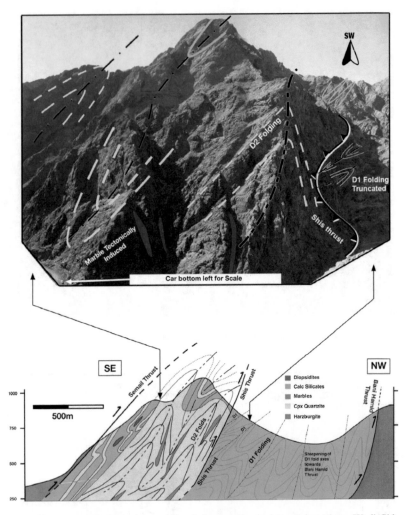

Fig. 7.14 Large-scale folds in granulite facies marbles and quartzites along Wadi Shis, Bani Hamid thrust sheet, photo courtesy Alan Cherry, in Searle et al. (2015)

the tectonic evolution of the Bani Hamid thrust slice in relation to obduction of the ophiolite is shown in Fig. 7.15.

The Bani Hamid granulites are interpreted as resulting from sub-ophiolite metamorphism of a larger, deeper slice of distal Tethyan oceanic rocks, probably

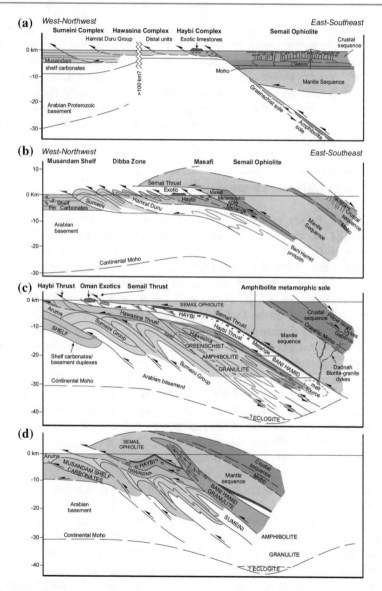

Fig. 7.15 Tectonic evolution of the Semail ophiolite and underlying thrust sheets during its formation **a** at 96–95 Ma, **b** at the Cenomanian–Turonian boundary, 94–93 Ma, when the metamorphic sole had accreted along the base of the ophiolite, **c** at 93–92 Ma when the Bani Hamid granulites reached peak metamorphic conditions in the lower crust, and **d** at 91–90 Ma showing final exhumation and thrusting of the Bani Hamid thrust sheet into the Semail ophiolite

similar to the Haybi complex during the ophiolite obduction process. The Madhah region shows excellent water worn exposures along the wadi. Access to Madhah is gained from the Fujairah–Khor Fakkan coastal road and a turning signposted to Madhah. A dirt road also winds down to Wadi Shis and ends at the tiny village of Shis with a perennial flowing wadi and small wadi pool and waterfall at the end. A dirt road winds from Wadi Shis through the ophiolite to the west towards Wadi Ham, and a new road is being constructed across the peridotite mountains to the west to connect with Masafi.

7.5   Ras Dadnah Leucogranite Dykes

The Khawr Fakkan ophiolite block shows numerous granitic dykes of varying mineralogy and composition intruding the ophiolite mantle sequence (Peters and Kamber 1994; Cox et al. 1999). A cluster of these dykes is present in Wadi Hulw bin Sulayman near the village of Bulaydah on the Masafi to Fujairah road. Another cluster of dykes and sills intrudes the mantle peridotites around Ras Dadnah on the east coast (Figs. 7.16 and 7.17), and the next small beach to the north (Fig. 7.18). Some composite sills at Ras Dadnah show perthitic intergrowths of K-feldspar and plagioclase, with the marginal facies rich in spessartine (Mn-rich) garnet, muscovite and tourmaline (Fig. 7.19a). Another prominent dyke complex at Dadnah has very large pink andalusite crystals with red garnet, white muscovite and black tourmaline, in addition to quartz and potassium feldspar (Fig. 7.19b). Similar andalusite pegmatites occur in a two meter wide dyke at Wadi Zikt, north of Dadnah (Fig. 7.20). At Ras Dibba a suite of more homogeneous pegmatite dykes contains abundant cordierite and andalusite with inclusions of staurolite and hercynite, rare biotite and secondary muscovite (Fig. 7.21). These granites crystallised at low pressures, between 3 and 5 kbar, by dehydration of muscovite-rich sediments. The depths are compatible with a melt source immediately beneath the ophiolite after it was thrust onto the Haybi complex and the underlying continental margin sediments.

The dykes appear to be restricted to the upper levels of the mantle sequence east of the Bani Hamid thrust sheet. A few dykes have hornblende diorite metaluminous compositions, but most are peraluminous and include biotite leucogranites, cordierite + andalusite + biotite   monzogranites,   garnet + tourmaline + muscovite + andalusite leucogranites. Rare topaz and lepidolite (lithium mica) occur in some dykes. Secondary muscovite formed at the expense of orthoclase, cordierite and andalusite. Single andalusite and tourmaline crystals can reach 12 cm in pegmatite dykes that are only one or two meter wide. These S-type granites

Fig. 7.16 Mountains around Ras Dadnah beach with upper mantle harzburgites and dunites and lower crustal gabbros; photo taken in 1979 prior to modern developments

Fig. 7.17 Ras Dadnah headland showing leucogranite dykes intruding harzburgites of the Mantle sequence

Fig. 7.18 **a** Dadnah cove, north of Ras Dadnah, showing leucogranite dykes intruding into the peridotites and gabbros of Moho transition zone. **b** Network of leucogranite dykes intruding mantle harzburgites and lower crust gabbros, Dadnah cove

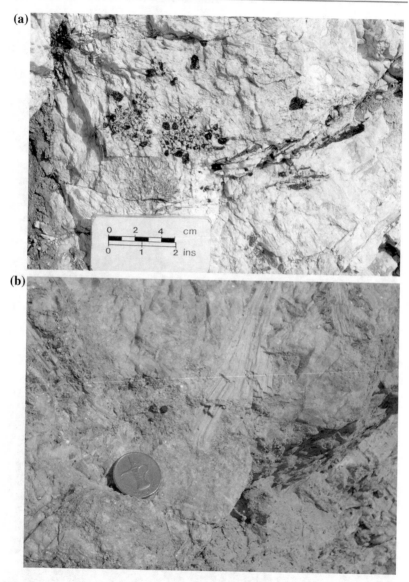

Fig. 7.19 a Large tourmaline (black) crystals, garnet (red) and andalusite (pink) in leucogranite from a dyke, Ras Dadnah. **b** Large pink andalusite crystals in leucogranite from a dyke intruding the upper mantle, Ras Dadnah

Fig. 7.20 Single tourmaline crystal 5 cm wide with red garnet and pink andalusite in a leucogranite dyke, Zikt, UAE

Fig. 7.21 Leucogranite dyke about 4 m wide intruding harzburgites of the mantle sequence, Ras Dibba, UAE

(derived from melting of sedimentary rocks) are interpreted to have been derived from melting a sedimentary source beneath the obducting ophiolite during the late stage of emplacement (Cox et al. 1999).

7.6 Khawr Fakkan Ophiolite

The Khawr Fakkan 'block' is the northern extent of the Oman–UAE ophiolite, a single thrust sheet that extends for over 600 km along the length of the mountains. The ophiolite in UAE shows a large extent of mantle rocks, mainly harzburgites (olivine + enstatite orthopyroxene) and dunite (olivine + minor chrome spinel), with a ~1 km thick Moho Transition zone (MTZ) overlain by lower crust gabbros, and sheeted dykes (Gnos and Nicolas 1996; Styles et al. 2006). The volcanic part of the ophiolite is exposed south of the border in Oman. The upper part of the mantle section is marked by a thick layer of dunite with complex intrusive relationships with late wehrlites (olivine + clinopyroxene ± orthopyroxene) and gabbros in the MTZ. The Moho is rarely a distinct mappable line, as in several places in Oman, but rather a zone of transition where gabbros become increasingly volumetrically important up-section (Fig. 7.22). There is debate whether extensive gabbros intrude through the mantle section (Styles et al. 2006; Goodenough et al. 2010) or whether the MTZ is gently folded and gabbros only occur above it in the crustal section (Ambrose and Searle 2019). The massive gabbros have cumulate textures in the lower parts showing early crystallization of mafic minerals and inter-cumulus plagioclase often reflecting batch melting along a spreading center. Dark-coloured wehrlites intrude the Moho Transition zone (Fig. 7.23). The cumulate gabbros pass upward to 'vari-textured' gabbros, and gabbros with a 'vinaigrette' texture showing rounded blocks of dark coloured diorite in a mass of felsic tonalites or trondhjemite (plagiogranite). These rocks can be interpreted either as reflecting coeval basic and felsic magma mingling as immiscible liquids, or as later intrusion of the pale tonalite breaking up blocks of older diorite.

Probably the best location to see all these ophiolite outcrops is along the recently constructed Sheik Khalifa bin Zayed road through the mountains to Fujairah where a series of new road cuts show spectacular pristine sections through key outcrops (Fig. 7.24). The mantle sequence peridotites extend a long way west as far as the base of Jebel Faiyah and other foreland jebels. The undulating Moho transition zone is gently folded in the west before dipping eastwards approaching Fujairah. Higher levels of the gabbroic lower crust sequence appear towards the east with sheeted dykes outcropping along the coast region north of the Oman border. The whole ophiolite section has been cut by the NW-SE striking Wadi

Fig. 7.22 Roadside outcrops along Sheikh Khalifa bin Zayed Highway to Fujairah showing dunites and gabbros in the Moho Transition zone, after Ambrose and Searle (2019)

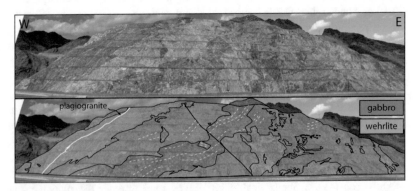

Fig. 7.23 Complex intrusive relationships between gabbros, wehrlite and plagiogranite, Sheikh Khalifa bin Zayed Highway, Fujairah, after Ambrose and Searle (2019)

Ham fault, a probably strike-slip fault with a minor component of normal motion, down-to-south. Several large quarries to the south of the road exploit massive gabbros for roadstone.

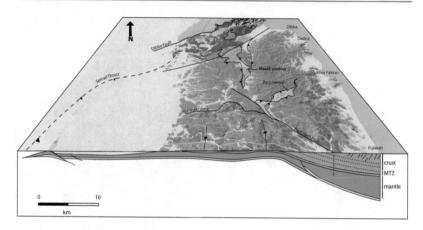

Fig. 7.24 Block diagram showing large-scale structure of the UAE section across the ophiolite, Fujairah

The structure and origin of the ophiolite continues to be controversial. One model suggests that the entire ophiolite formed along a fast spreading Mid-Ocean ridge (MOR) similar to the East Pacific rise (Nicolas et al. 2000; Nicolas and Boudier 2011). An alternative model suggests two-stage magmatism with Phase 1 related to normal MOR magmatism and the volumetrically greater Phase 2 related to supra-subduction zone magmatism (Styles et al. 2006; Goodenough et al. 2014). A third model relates the entire ophiolite as forming above a NE-dipping subduction zone (Pearce et al. 1981; Searle and Cox 1999, 2002; Searle et al. 2003, 2015; MacLeod et al. 2013). Combining all the structural, metamorphic, geochemical and geochronological data strongly suggests that the supra-subduction origin of the whole ophiolite.

7.7 Wadi Hatta

Wadi Hatta is a scenic wadi that cuts through the entire mountains crossing the UAE–Oman border. A main road connecting Madam in UAE with Shinas and the Batinah coast in Oman runs along Wadi Hatta. Jebel Raudha at the western end of Wadi Hatta in the northern enclave of Oman, is an impressive peak showing important structural relations between the ophiolite and Hawasina series with overlying Maastrichtian and Paleocene limestones (Figs. 7.25 and 7.26). It is possible to trek along the whole summit ridge of Jebel Raudha and see the folded ophiolite sequence, the impressive unconformity above Hawasina complex cherts

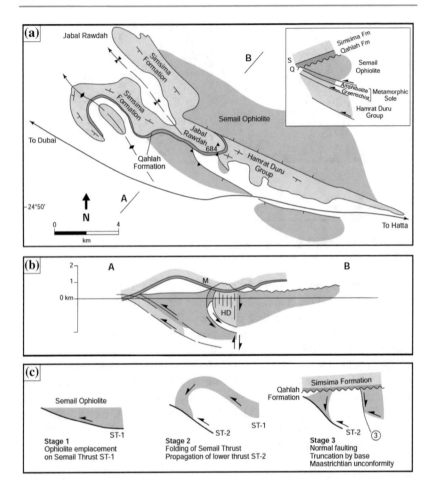

Fig. 7.25 Geological map of Jebel Raudha, UAE–Oman border

and the folded Paleogene limestones at the far western end of the mountain. Structural mapping of Jebel Raudha has revealed three stages in the evolution of this mountain belt, with the initial shallow-dipping Semail thrust folded and cut by later breakback thrusting, with all structures truncated by the Maastrichtian unconformity along the base of the Qahlah and Simsima Formations. Later more gentle folding has also affected the Maastrichtian—Paleogene sedimentary rocks. Onlapping bedding in the Simsima Formation limestones suggest Late Cretaceous paleo-topography to the west.

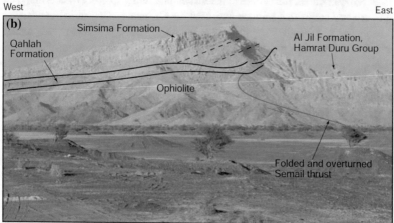

Fig. 7.26 Field photographs of Jebel Raudha showing tightly folded Semail thrust with ophiolite above highly contorted Hawasina cherts of the Al Jil Formation

A dirt road runs south connecting Wadi Hatta with the northern end of Jebel Sumeini, but border controls have now restricted access to this route. Further east along Wadi Hatta the road crosses back into UAE territory. A series of beautiful blue wadi pools can be found along an incised gorge cut through the ophiolite near another dirt road running south of Wadi Hatta to Wadi Rayy. This road winds through spiky black mountains composed of mantle sequence harzburgites and eventually connects through to the Omani village of Shuwayhah and the Sumeini Window (see Chap. 8). East of Hatta a second border crossing into Oman leads on to the upper sections of the ophiolite in Oman territory.

7.8   Jebel Faiyah–Jebel Maleihah

Several small jebels along the western flank of the mountains show Maastrichtian and Paleocene limestones unconformably overlaying serpentinised peridotites of the leading edge of the Semail ophiolite thrust sheet. The base of the neo-autochthonous sequence unconformably above the ophiolite is marked by the Qahlah Formation comprising reddish siltstones and conglomerates that contain abundant fossil fragments as well as peridotite pebbles. An abrupt transition to shallow-marine limestones is apparent at the base of the Simsima Formation. Upper Maastrichtian limestones contain abundant rudists (*Durania*, *Dictyoptychus*, sp.), solitary corals that became extinct at the Cretaceous-Cenozoic boundary and *Acteonellid* gastropods. Structural style is one of fault-propagation folds, usually with box-folding geometry. In many localities another unconformity above the Maastrichtian limestones suggest differential uplift after obduction. Paleocene limestones of the Malaqet and Mundasa (equivalent to the Umm er Radhuma) Formations are common on many of the foreland jebels. Some localities (Jebel Auha, Jebel Huwayyah) show Middle Eocene Dammam Formation limestones overlying Maastrichtian Simsima Formation.

7.9   Jebel Hafit Anticline

Jebel Hafit is a spectacular Cenozoic 'whaleback' pericline fold approximately 22 km long and more than 4 km wide, south of the city of Al Ain (UAE) and Buraimi (Oman). It is an asymmetric ENE-verging, doubly plunging anticline with a steep, even slightly overturned, eastern limb and a gentle dipping west limb (Fig. 7.27). It has a rounded box-fold shape with a minor east-vergent thrust (Tarabat thrust) along the eastern kink band set. The main Hafit anticline axis plunges NNW and a subsidiary anticline splays off the northeast (Al-Ain anticline) with an asymmetric syncline (Rwaidhat syncline) separating the two anticline axes (Fig. 7.28). The east-vergent Jebel Hafit fold is antithetic to the dominant WSW

Fig. 7.27 Aerial photograph of the Jebel Hafit pericline in Cenozoic limestones, Al Ayn (UAE)–Buraimi (Oman) foreland region west of the mountains. Photo taken in 1978 prior to major developments around Al Ayn

vergence of other folds and the WSW transport of thrust sheets in the northern mountains.

The oldest rocks exposed in the core of the anticline are Lower Eocene Rus Formation limestones with successive Middle-Upper Eocene Dammam and Lower-Middle Oligocene Asmari Formation limestones above. Balanced cross-sections indicate that the Jebel Hafit fold is cored by Aruma Group shales at depth with a probable detachment along the top of the Cretaceous shelf carbonates (Fig. 7.28). The youngest rocks affected by the folding are Miocene Fars Formation limestones and gypsum clays, although the dips in the Fars beds are not as steep as the Eocene–Oligocene rocks. Searle and Ali (2009) show more detailed maps and seismic sections running across Jebel Hafit and the foreland west of the ophiolite in this district.

The Jebel Hafit anticline appears similar to many of the classic Zagros folds in Iran and must have formed after the Miocene, a similar age of deformation as those in the Zagros Mountains. There is some evidence that the fold started to grow during the Paleogene and continued into the Miocene. Seismic profiles across the Oman–UAE foreland show Cenozoic folding throughout, but few examples are

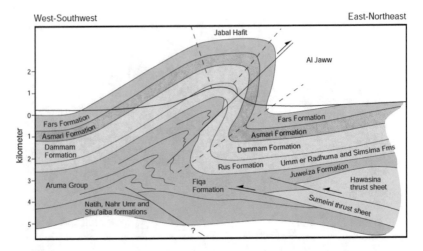

Fig. 7.28 Cross-section of the Jebel Hafit anticline showing stratigraphy and east-vergence of the large-scale pericline fold. Lower structure is constrained by seismic profiles (Searle and Ali 2009)

uplifted and exposed as much as Jebel Hafit making this an important geological site (Searle and Ali 2009). Major developments in the UAE part of Jebel Hafit around Al Ain have spread around the jebel, but the Oman part of the mountain remains in a pristine state. Thesiger (1975) described hunting for Arabian tahr (mountain goats) on Jebel Hafit in the 1940s and it is still possible that a few specimens remain. The southern part of the Jebel Hafit anticline has recently been drilled for oil, without success.

7.10 Liwa Oasis

Liwa oasis is about 90 km long and thought to be largest oasis in Arabia, if not the World. The Liwa district of southern Abu Dhabi has some of the largest sand dunes in UAE along the border with Saudi Arabia (Fig. 7.29). Individual barchan type dunes are over 150 m above the inter-dune gypsum-cemented sabkhas and were built by winds blowing dominantly from the SSE (Glennie 2005). The base level is about 100 m above sea-level. The largest dunes are flanked to the NW by smaller transverse dunes (*aklé*) that can migrate rapidly, engulfing larger dunes. In the southwest Liwa oasis transverse dunes have been modified by winds from the northwest to give their pattern a linear aspect. Many dunes have slipfaces over

Fig. 7.29   Sand dunes in the Liwa oasis, UAE

100 m high with spectacular avalanche slopes continually resulting in dune migration. In general, transverse dunes result from wind systems blowing from the north (*shamal*) and linear dunes from winds blowing from the west. Many shifts in orientation and strength of winds must have occurred both annually and over longer time spans. Along the north and west margins of the Liwa oasis the mega dunes and sabkhas are covered by sand derived from the northwest. In the main dune fields water is at well depth in the Liwa oasis so there is no surface evaporation to create sabkhas.

References

Ambrose TK, Wallis D, Hansen LN, Waters DJ, Searle MP (2018) Controls on the rheological properties of peridotite at a palaeosubduction interface: a transect across the base of the Oman-UAE ophiolite. Earth Planet Sci Lett 491:193–206

Ambrose TK, Searle MP (2019) 3-D structure of the Northern Oman–UAE ophiolite: widespread, short-lived, suprasubduction zone magmatism. Tectonics 38. https://doi.org/10.1029/2018TC005038

Cowan RJ, Searle MP, Waters DJ (2014) Structure of the metamorphic sole to the Oman Ophiolite, Sumeini Window and Wadi Tayyin: implications for ophiolite obduction processes. In: Rollinson HR, Searle MP, Abbasi IA, Al-Lazki A, Al-Kindy MH (eds) Tectonic evolution of the Oman mountains. Geological Society, London, Special Publication, vol 392, pp 155–176

Chauvet F, Dumont T, Basile C (2009) Structures and timing of Permian rifting in the central Oman Mountains (Saih Hatat). Tectonophysics 475–574

Cox JS, Searle MP, Pedersen RB (1999) The petrogenesis of leucogranite dykes intruding the northern Semail ophiolite, United Arab Emirates: field relations, geochemistry and Sr/Nd isotope systematics. Contrib Miner Petrol 137:267–287

Fortey RA, Heward AP, Miller CG (2011) Sedimentary facies and trilobite and conodont faunas of the Ordovician Rann Formation, Ras al Khaimah, United Arab Emirates. GeoArabia 16:127–152

Glennie KW (2005) The deserts of Southeast Arabia. GeoArabia, Manama, Bahrain, 215 p

Gnos E (1998) Peak metamorphic conditions in garnet amphibolites beneath the Semail ophiolite: implications for an inverted pressure gradient. Int Geol Rev 40:281–304

Gnos E, Nicolas A (1996) Structural evolution of the northern end of the Oman Ophiolite and enclosed granulites. Tectonophysics 254:111–137

Goodenough KM, Styles M, Schofield DI, Thomas RJ, Crowley Q, Lilly RM et al (2010) Architecture of the Oman–UAE ophiolite: evidence for multi-phase magmatic history. Arab J Geosci 3:439–458

Goodenough KM, Thomas RJ, Styles M, Schofield DI, MacLeod CJ (2014) Records of ocean growth and destruction in the Oman–UAE ophiolite. Elements 10(2):109–114

Maury R, Bechennec F, Cotton J, Caroff M, Cordey F, Marcoux J (2003) Middle Permian plume-related magmatism of the Hawasina Nappes and the Arabian Platform: implications on the evolution of the Neotethyan margin in Oman. Tectonics 22:1073. https://doi.org/10.1029/2002TC001483

MacLeod C, Lissenberg J, Bibby LE (2013) 'Moist MORB' axial magmatism in the Oman ophiolite: the evidence against a mid-ocean ridge origin. Geology 41:459–462

Nicolas A, Boudier F (2011) Structure and dynamics of ridge axial melt lenses in the Oman ophiolite. J Geophys Res 116:B03103

Nicolas A, Boudier F, Ildefonse B, Ball E (2000) Accretion of Oman and United Arab Emirates ophiolite—discussion of a new structural map. Marine Geophys Res 21:147–180

Pearce JA, Alabaster T, Shelton AW, Searle MP (1981) The Oman ophiolite as a Cretaceous arc-basin complex: evidence and implications. Philos Trans R Soc Lond A300:299–317

Peters T, Kamber B (1994) Peraluminous potassium-rich granitoids in the Semail ophiolite. Contrib Miner Petrol 118:229–238

Pillevuit A, Marcoux J, Stampfli G, Baud A (1997) The Oman Exotics: a key to the understanding of the Neotethyan geodynamic evolution. Geodin Acta 10(5):209–238

Rioux M, Garber J, Bauer A, Bowring S, Searle MP, Keleman P, Hacker B (2016) Synchronous formation of the metamorphic sole and igneous crust of the Semail ophiolite: new constraints on the tectonic evolution during ophiolite formation from high-precision U–Pb zircon geochronology. Earth Planet Sci Lett 451:185–195

Searle MP, Ali MY (2009) Structural and tectonic evolution of the Jabal Sumeini–Al Ain–Buraimi region, northern Oman and eastern United Arab Emirates. GeoArabia 14:115–142

Searle MP, Cox JS (1999) Tectonic setting, origin and obduction of the Oman ophiolite. Geol Soc Am Bull 111:104–122

Searle MP, Cox JS (2002) Subduction zone metamorphism during formation and emplacement of the Semail Ophiolite in the Oman Mountains. Geol Mag 139:241–255

Searle MP, Malpas J (1980) The structure and metamorphism of rocks beneath the Semail Ophiolite of Oman and their significance in ophiolite obduction. Trans R Soc Edinb Earth Sci 71:247–262

Searle MP, Malpas J (1982) Petrochemistry and origin of sub-ophiolite metamorphic and related rocks in the Oman Mountains. J Geol Soc Lond 139:235–248

Searle MP, Graham GM (1982) "Oman Exotics"—oceanic carbonate build-ups associated with the early stages of continental rifting. Geology 10:43–49

Searle MP, Cherry AG, Ali MY, Cooper DJW (2014) Tectonics of the Musandam Peninsula and northern Oman Mountains: from ophiolite obduction to continental collision. GeoArabia 19:135–174

Searle MP, Warren CJ, Waters DJ, Parrish RR (2003) Subduction zone polarity in the Oman mountains: implications for ophiolite empplacement. In: Dilek Y, Robinson PT (eds) Ophiolites in earth history. Geological Society, LondonSpecial Publication 218, pp 467–480

Searle MP, Waters DJ, Garber JM, Rioux M, Cherry AG, Ambrose TK (2015) Structure and metamorphism beneath the obducting Oman ophiolite: evidence from the Bani Hamid granulites, northern Oman Mountains. Geosphere 11(6). https://doi.org/10.1130/ges01199.1

Searle MP, James NP, Calon TJ, Smewing JD (1983) Sedimentological and structural evolution of Arabian continental margin in the Musandam Mountains and Dibba zone, United Arab Emirates. Geol Soc Am Bull 94:1381–1400

Soret M, Agard P, Dubacq B, Plunder A, Yamato P (2017) Petrological evidence for stepwise accretion of metamorphic soles during subduction infancy (Semail ophiolite, Oman and UAE). J Metamorph Geol 35:1051–1080

Styles M, Ellison RA, Phillips ER et al (2006) The geology and geophysics of the United Arab Emirates, vol 2. Ministry of Energy, United Arab Emirates, Abu Dhabi

Thesiger W (1975) Arabian Sands. Longmans

Northern Oman Mountains

8

The northern part of the Oman Mountains includes the largest extent, best exposures and most complete sections through the ca 15–20 km thick Semail ophiolite (Fig. 8.1). Many of these geological sites are of World-class significance and a Semail Ophiolite GeoPark has been proposed covering the best exposures around Wadi Jizzi. This chapter covers the mountains south of the UAE border and north of the Hawasina Window and Jebel al-Akhdar. The ophiolite exposures in northern Oman include the full range of volcanic rocks that are not exposed in the UAE to the north. These include the classic 'Geotimes' (V1) pillow lavas beautifully exposed along Wadi Jizzi and the important 'Lasail' volcanics (V2) which are largely boninitic in composition and show evidence for the Supra-subduction zone origin of the lavas. The ancient copper mines of Lasail and Aarja are located around the eastern end of Wadi Jizzi, and the most extensive chromite mines come from the high mountains beneath the Moho in the Wadi Rajmi and Wadi Fizh area, north of Wadi Jizzi.

Apart from prime locations and sites within the ophiolite, this area includes some of best fossil sites in the Oman–UAE Mountains at Jebel Sumeini and the fold and thrust structure within the Sumeini Group. The Sumeini Window, where erosion has exposed a complete section through the base of the ophiolite shows spectacular exposures of the metamorphic sole and the thrust structures in the Haybi and Hawasina complex beneath. Some important Upper Triassic 'exotic' limestones crop out at Jebel Ghawil, south of the Sumeini area. This mountain massif shows impressive thrusts and duplex structures repeating the coral limestones and underlying Haybi volcanics, with some very unusual highly alkaline intrusions of gabbro and ultramafic rocks indicative of Triassic ocean islands.

© Springer Nature Switzerland AG 2019
M. Searle, *Geology of the Oman Mountains, Eastern Arabia*,
GeoGuide, https://doi.org/10.1007/978-3-030-18453-7_8

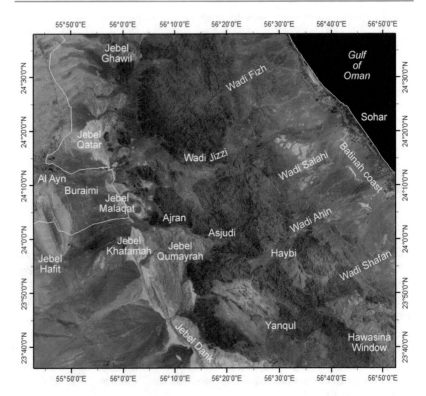

Fig. 8.1 Landsat photo of northern Oman showing key locations

8.1   Jebel Sumeini North

Jebel Sumeini in the northern part of the Oman Mountains forms two large-scale west-facing folds in the proximal shelf edge limestones of the Sumeini Group (Figs. 8.2 and 8.3; Glennie et al. 1973, 1974; Watts and Garrison 1986; Watts 1990). At the far northwestern end of the jebel a major unconformity dips to the west above which Upper Maastrichtian Simsima Formation limestones contains up to 90% beautifully preserved coiled *Acteonellid* gastropods and rudists (*Durania, Dictyoptychus, Hippurites cornucopia*) in a 'death assemblage' (Fig. 8.4a–c; Skelton et al. 1990; Searle and Ali 2009). The structural relationships at Jebel Sumeini also provide clear evidence of the timing of the ending of Late Cretaceous

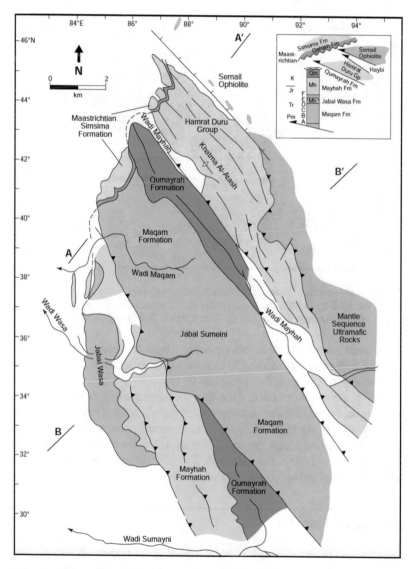

Fig. 8.2 Geological map of the Jebel Sumeini area, northern Oman, with key structural features shown in inset

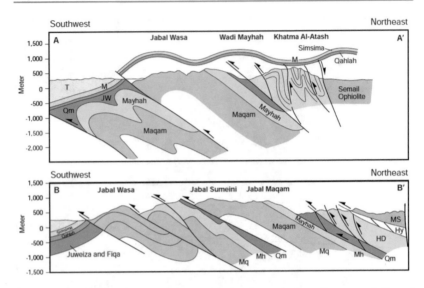

Fig. 8.3 Cross-sections across Jebel Sumeini showing main Late Cretaceous folds and thrusts truncated by the Maastrichtian–Danian unconformity. The thrust along the western margin of Jebel Sumeini cuts up into the Maastrichtian Qahlah and Simsima Formations

thrust emplacement of the Semail Ophiolite, Haybi and Hawasina thrust sheets onto the Arabian passive margin (70 Ma) and a subsequent phase of folding during the Danian period of Early Paleogene (65–63 Ma). The gastropods and rudists are in many places weathered out of the rock so samples are easily collectable, but this site is an important SSSI site, so that no collecting of any fossil material is allowed. Easy access from the UAE has meant that considerable destruction of the site has already occurred, although the newly constructed fence along the Oman–UAE border has relieved this to some extent. It is crucial that this World-class fossil site is fenced off and protected as a GeoPark–SSSI site.

8.2   Jebel Sumeini

Jebel Sumeini is entirely within Oman just south of the UAE border. It comprises three doubly-plunging folds above thrust faults that strike NNW-SSE (Fig. 8.5). The structurally higher eastern Jebel Sumeini is composed of Middle Permian to Early Jurassic Maqam Formation shelf margin with the stratigraphically higher Middle Jurassic to Late Cretaceous Mayhah, Huwar and Qumayrah Formations

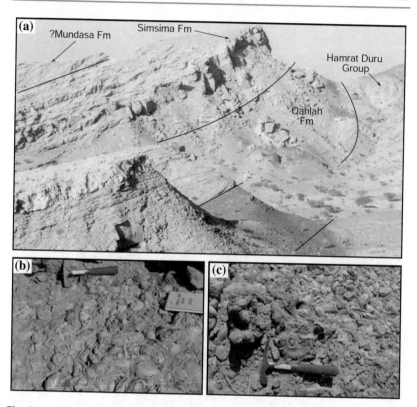

Fig. 8.4 **a** Northern corner of Jebel Sumeini with Maastrichtian Qahlah and Simsima Formations unconformably overlying the folded Triasssic-Jurassic limestones of Jebel Sumeini. **b** and **c** Death assemblage fossils including rudists, corals and gastropods in the Maastrichtian Qahlah and Simsima Formations

outcropping only in the north along the western side of Wadi Mayhah (Fig. 8.6; Cooper et al. 2018). The western fold shows the Jebel Wasa Formation, a Late Triassic facies of the Sumeini Group comprising calcirudites with abundant coral and algal debris interpreted as a fore-reef slope setting. This margin of Jebel Wasa also shows important key outcrops of gypsiferous sediments derived from the Precambrian Ara salt (Fig. 8.7). The eastern margin of Jebel Sumeini may be a normal fault with structurally overlying Hamrat Duru Group (Hawasina complex) rocks exposed to the east. Folds are west-verging and are the latest in the succession of ophiolite obduction related structures (Fig. 8.8). The folded and faulted

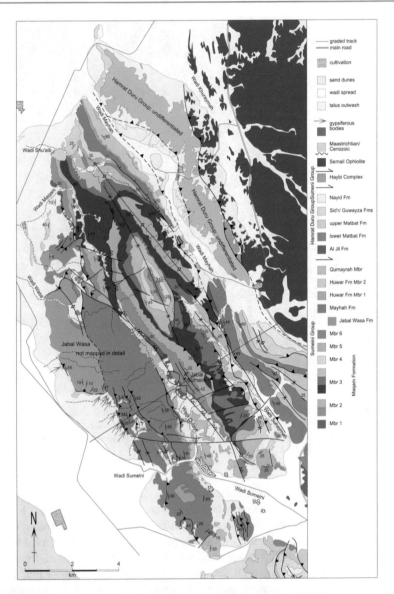

Fig. 8.5 Geological map of Jebel Sumeini area, after Cooper et al. (2012)

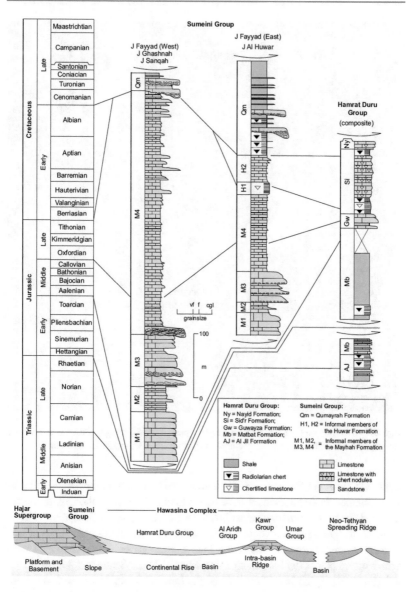

Fig. 8.6 Stratigraphic log sections through the Sumeini Group sediments, and Hamrat Duru Group, after Cooper et al. (2012)

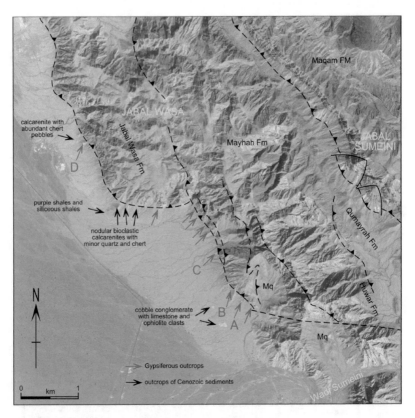

Fig. 8.7 Satellite photo of Jebel Wasa, western part of Jebel Sumeini, showing locations of the gypsiferous sediments probably intruded upward from Precambrian Ara salt

structures on Jebel Sumeini are all truncated and overlain unconformably by the Late Campanian–Maastrichtian terrestrial clastic sediments of the Qahlah Formation and the shallow-marine rudist limestones of the Simsima Formation (Skelton et al. 1990). The structural geometry shows that the major folds and thrusts in Jebel Sumeini were all formed during the Late Cretaceous ophiolite obduction event which ended before the Late Campanian–Maastrichtian (Searle and Ali 2009). The structures underlying Jebel Sumeini and the foreland region to the west are shown in the seismic interpretation of Fig. 8.9. Both the Qahlah and Simsima Formation limestones with their 'death assemblage' rudists and gastropods now dip gently to the west.

Fig. 8.8 West-vergent folds in the Sumeini Group carbonates, Jebel Sumeini

Along the central western part of Jebel Sumeini, a few scattered outcrops of gypsum–anhydrite with rafts of creamy brown well-bedded sandstone and silt-stone, micaceous quartzites and pink-grey fine-grained limestone are exposed (Cooper et al. 2018). A cobble conglomerate also contains abundant clasts of sandstones and limestones. The gypsiferous bodies form a band about 4 km long and up to 100 m wide along a late-stage thrust that re-thrusted the previous allochthonous section over the younger Maastrichtian-Paleogene cover rocks. High Sr isotope ratios suggest correlation with the Ediacaran–Early Cambrian salt domes of central Oman and their emplacement is thought to have been along deep seated faults, similar to the salt-anhydrite-gypsum deposits in Jebel Qumayrah and the Hawasina Window (Chap. 9). The Jebel Sumeini gypsum-anhydrite is close to the Hormuz (Ara) salt basin of UAE and the Arabian Gulf area, and is of similar age to the Fahud and Ghaba salt basins in central and eastern Oman.

The Jebel Sumeini limestones are surrounded by the Semail ophiolite thrust sheet which has been folded around the later culmination of Jebel Sumeini (Fig. 8.10). A series of jebels showing the overlying Paleocene-Eocene limestones outcrop along the western margin of mountains south of Jebel Sumeini (Fig. 8.11). These show gentle west-verging folding along the western margin exposed at Sulayf (Fig. 8.12) for example.

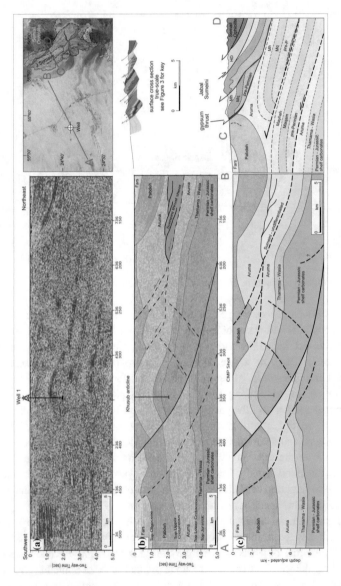

Fig. 8.9 Sub-surface structure of the area west of Jebel Sumeini from seismic sections, after Searle and Ali (2009)

Fig. 8.10 Landsat photo of the northern Oman mountains showing locations of the proposed GeoPark sites

8.3   Sumeini Window Metamorphic Sole

The best exposures of the sub-ophiolite metamorphic sole in Oman lie in the northwest corner of the Sumeini Window (Fig. 8.13; Searle and Malpas 1980, 1982; Searle and Ali 2009; Cowan et al. 2014). Thrust slices of gar- net + clinopyroxene granulites and amphibolites underlie the 'banded ultramafic unit' (harzburgites, lherzolites, dunites) along the base of the ophiolite mantle sequence (Fig. 8.14). The olivine and orthopyroxene-bearing harzburgites contain a minor amount of clinopyroxene (chrome diopside) and up to 10–20% horn- blende, which shows that metamorphism affected the base of the ophiolite as well

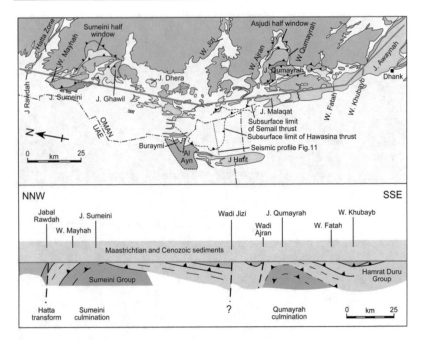

Fig. 8.11   Geological map and north-south section along the foreland jebels south of Jebel Sumeini, after Cooper et al. (2012)

as the sole amphibolites and greenschists beneath the Semail thrust (Fig. 8.15). Individual garnets can be large with halos of clinopyroxene surrounded by horn-blende + plagioclase assemblages (Fig. 8.16). Pressure-temperature conditions from the garnet + clinopyroxene granulites immediately beneath the Semail thrust reveal conditions of formation around 770–900 °C and 11–13 kbar (Cowan et al. 2014). The rocks are highly deformed showing mylonitic textures and the thermal profile shows that metamorphism is highly condensed and inverted with higher grade rocks sheared above lower grade rocks. The peak pressures suggest that the depth of metamorphism ($\sim$45–55 km) was more than double the thickness of the preserved ophiolite ($\sim$15–20 km), so the heat source must have been the mantle wedge above the subduction zone. U-Pb zircon ages of partial melt pods within the highest temperature parts of the sole (96.16 Ma; are almost identical to the ages of the ophiolite (96.12–95.50 Ma; Rioux et al. 2016). The synchroneity of ophiolite crust and metamorphic sole formation can only point to a supra-subduction zone setting for ophiolite formation and initiation of subduction/obduction.

Fig. 8.12 Sulayf fort, built on Paleocene limestones with the frontal fold of the Ibri jebel in the background

Beneath the amphibolites imbricated slices of greenschist facies marbles, pink piemontite quartzites derived from distal oceanic Mn-rich cherts (Fig. 8.17), and rare pelitic rocks (Haybi complex thrust sheet) are thrust above unmetamorphosed rocks of the Hamrat Duru Group and other allochthonous Hawasina complex thrust sheets (Fig. 8.18). The metamorphic sequence is inverted with higher grade rocks thrust above lower grade and the entire sequence shows a strong mylonitic fabric. The Sumeini Window sole rocks are crucially important for interpretation of the obduction history of the Oman ophiolite. These rocks are probably the most intact and complete section of sub-ophiolite metamorphic sole anywhere in the World. The best locations are the Shuwayhah gorge (Fig. 8.19) where the full condensed section of granulites, amphibolites and greenschists is exposed, and the northern margin of the Window where prominent Banded ultramafic unit peridotites overlie garnet + clinopyroxene amphibolites and granulites. The amphibolites and greenschist facies rocks of the metamorphic sole have been imbricated in a series of thrust slices immediately beneath the Semail Thrust (Fig. 8.20). Access is tricky with a maze of four-wheel drive dirt tracks up several wadis; the advice is simply to head north and aim for the large ophiolite mountains.

Fig. 8.13 Metamorphic sole beneath the Semail ophiolite in the Sumeini Window, Dark hills of harzburgite mantle sequence rocks overlie the olive green coloured amphibolites of the upper metamorphic sole, in turn overlying white marbles and greenschist facies meta-sediments

8.4 Jebel Ghawil Exotic

Jebel Ghawil is an Upper Triassic 'exotic' limestone mountain located south of the Sumeini Window immediately beneath the ophiolite and its imbricated metamorphic sole (Fig. 8.21). Access is by 4-wheel drive from the Wadi Jizzi–Buraimi road north to Mahadah along the western margin of the mountains. From Mahadah head north towards Sumeini and Jebel Ghawil is the prominent white exotic limestone mountain to the west of the road before the Shuwayah turn-off. Jebel Ghawil is important for several reasons, firstly for its classic exposures of thrust repeated units of Upper Triassic Exotic limestones containing the large bivalve *Megalodon*

Fig. 8.14 Banded ultramafic unit comprising dark harzburgites and pale brown dunites, with strong shearing fabrics along the Semail thrust, base of the ophiolite

sp. overlying ankaramites and alkali basalts of the Haybi complex (Searle et al. 1980), and secondly for the presence of some remarkable differentiated alkaline ultrabasic sills within the Upper Triassic volcanic sequence (Searle 1984). The largest sill consists of jacupirangite (kaersutite amphibole + titanaugite clinopyroxene enclosed in large poikilitic biotite) at the base and kaersutite gabbro at the top intruded into Haybi volcanic rocks. Two more sills 4 m thick exposed on Jebel Ghawil are composed of alkaline wehrlites (titanaugite + olivine + biotite). These unusual rock types are interpreted as off-axis intrusions into ocean islands that formed in a distal location within Tethys prior to the ophiolite emplacement process. Similar alkaline ultramafic rocks are found in ocean islands such as the Comores, Réunion and Tristan da Cunha in off-axis, within plate tectonic settings.

Ol + Opx ± Cpx
harzburgites

Grt + Hbl + Pl ± Cpx amphibolites
P-T ~800-880°C;11.5 kbar; ~44 km depth

Semail Thrust

Fig. 8.15 The Semail Thrust, a rare outcrop showing a brittle thrust fault with clinopyroxene-bearing harzburgite overlying garnet + diopside amphibolite of the Metamorphic sole, Shuwayhah gorge, Sumeini Window

8.5   Wadi Rajmi–Wadi Fizh Ophiolite

Wadi Rajmi and Wadi Fizh show particularly spectacular profiles across the northern part of the ophiolite. Access is from the Batinah coast road turning west towards the mountains south of Shinas. The Geotimes or V1 volcanic unit is up 900 m thick in this area, and shows classic pillow lavas, pahoehoe and lobate sheet flows with occasional hyaloclastites and hydrothermal breccias (Miyashita et al. 2012). Red and brown metalliferous umbers are intercalated with the lavas with bright red jaspers resulting from silica hydrothermal metasomatism. North-south trending sheeted dykes are particularly well exposed in these wadis with plagiogranite and tonalite intrusions along the root zone. In Wadi Fizh an east-west trending boninite dyke intrudes the sheeted dyke complex. The lower crust section is marked by massive gabbros in the higher part, and gabbronorite sills in the lower part. Dark coloured wehrlite intrusions cut the gabbronorites and all these rocks are cut by a dolerite dyke swarm. The Moho transition zone is only about 10 m thick

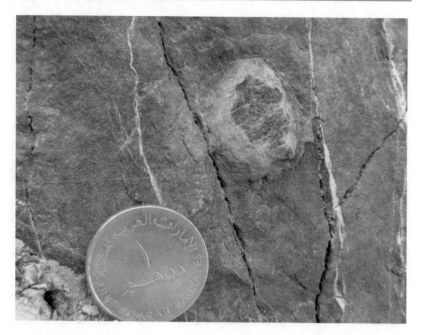

Fig. 8.16 Large garnets with clinopyroxene in granulite facies enclaves within amphibolites, Shuwayhah gorge, Sumeini Window

in Wadi Fizh, with orange–brown coloured dunites and darker wehrlites below the Moho (Kaneko et al. 2014), and gabbros and olivine gabbronorites above the Moho. In this region the Moho is vertical and has been interpreted as a sinistral ductile shear zone (Macleod et al. 2013)

In the Wadi Fizh–Rajmi region several small chromite deposits have been found. Chromite pods and lenses represent the most refractory phase of the mantle and seem to be located immediately beneath the Moho. In Wadi Rajmi some very rare examples of 'polo chromite' have been discovered. These are spherical nodules of black chromite enclosing a core of orange-coloured serpentinised dunite. Recently, chromite lenses in ophiolites from Tibet have revealed minute inclusions of extreme high-pressure mineral inclusions including phases such as coesite (high pressure polymorph of quartz), micro-diamond and minerals originating from lower mantle depths. It is possible that similar micron-scale inclusions are present in Omani chromites, but this has yet to be studied in detail. These chromite deposits are too small to be of any commercial value but are geologically very important.

Fig. 8.17 Greenschist facies metamorphic sole comprising white marbles and brownish quartzites, Shuwayhah gorge, Sumeini Window

8.6 Wadi Jizzi Ophiolite

The northern part of the Oman ophiolite is the most complete and best-exposed section of any ophiolite anywhere in the World. The proposed GeoPark site covers the area from Wadi Jizzi south to Wadi Salahi–Wadi Ahin and through the entire upper mantle and crust section through the complex (Fig. 8.22). Access is along the main road that connects Buraimi with Sohar, and many of the wadis to the south which can also be accessed by four-wheel drive from the Batinah coast. North of Wadi Jizzi the same ophiolite units continue to the Wadi Rajmi and Wadi Fizh regions and are exposed in the mountainous country up to the UAE border.

This GeoPark site contains several iconic outcrops of all components of the ophiolite including the Wadi Jizzi Geotimes pillow lavas (Fig. 8.23). These classic pillow lavas show dykes intruding beneath, long sinuous pillows indicative of eruption on a slope, and good way-up criteria. They are part of the Geotimes unit

Fig. 8.18 Tight–isoclinal folding in unmetamorphosed cherts and turbidites of the distal Hawasina complex immediately beneath the metamorphic sole

(because they appeared on the front cover of '*Geotimes*' magazine vol. 20 (8), courtesy of Bob Coleman). The Geotimes V1 lavas were initially thought to represent MORB, but trace element geochemistry and isotope data show that they are more evolved than MORB and relate to supra-subduction lavas (Pearce et al. 1981). This area also shows the classic outcrops of the Lasail arc volcanic unit. These lavas have an island arc boninite geochemistry and were instrumental in the first proposal of formation of the ophiolite in an island-arc setting (Pearce et al. 1981; Alabaster et al. 1982; Lippard et al. 1986). Lasail unit lavas are typically non-vesicular, porphyritic olivine ± clinopyroxene ± plagioclase phyric basalts and form small classic pillows. The Lasail pillow lavas are cut by a large number of orangy-brown andesitic cone sheets in distinct volcanic centres. The final phase of magmatism is a series of acid felsite sills that cut all previous magmatic units. Thin horizons of dark red to purple coloured umbers, shales and red radiolarian cherts are exposed between the Geotimes (V1) lavas and the Lasail (V2) arc lavas. Radiolaria from these cherts are Cenomanian-Turonian in age, exactly the same age as the U-Pb zircon ages from the ophiolite (96.5–95.5 Ma: Rioux et al. 2016).

Fig. 8.19 Amphibolites of the metamorphic sole overlying greenschist facies marbles, quartzites and rare meta-volcanic rocks, Shuwayha gorge, Sumeini

Also exposed in this area are rift-related Salahi and Alley units of clinopyroxene-phyric lavas.

Sheeted dykes are exposed along the whole mountain belt beneath the pillow lavas in northern Oman (Fig. 8.24). One of the most impressive outcrops is at Zabyat on the west margin of the 'Alley' north of Wadi Jizzi (Smewing 1981; MacLeod et al. 2013). Wadi Salahi and Wadi Suhayli, south of Wadi Jizzi contain some of the most complete and best sections of sheeted dykes up to 1.5 km thick anywhere in the World (Umino et al. 2003). Dykes are between a few cm and 13 m wide and commonly have chilled margins. They can be seen emanating from lower gabbros and feeding pillow lavas at the top in several spectacular outcrops. Some of these outcrops show 100% dykes indicating an extensional setting along an oceanic spreading ridge. Chilled margins are common and some later dykes can be seen cross-cutting earlier dykes. Structurally below the sheeted dykes, massive

Fig. 8.20 Imbricate thrust slices of the metamorphic sole marbles (white), meta-volcanics (dark) and amphibolites, Sumeini Window

gabbros occur in the mountains to the west. In the high mountains massive isotropic gabbros pass down to layered gabbros with increasing peridotite remnants in the Moho transition zone. The lower gabbros have cumulate textures and are interpreted as base of magma chamber igneous rocks, with batch melting continually feeding melts up from the upper mantle, and melt extraction along the upper gabbro sills and sheeted dykes above. The Moho is spectacularly exposed in the Wadi Jizzi area (Fig. 8.25).

8.7 Aarja, Bayda and Lasail Copper Mines

Important copper deposits at Lasail, Aarja and Bayda are associated with the arc volcanics and were mined by the Sumerians as long ago as the 3rd century BC. More recent mining in the 1990s has ceased and relic mine-shaft and open pits are left at Aarja site north of Wadi Jizzi. At Aarja the massive sulphide deposit was

Fig. 8.21 Jebel Ghawil Exotic, showing thrust slices of white Upper Triassic limestones, and dark Haybi alkali volcanics

originally about 75 m length and capped by a NW-trending orangy-brown coloured gossan. The deposit is located above the Geotimes V1 volcanic unit and along the base of the Lasail V2 lavas. A large plagiogranite intrudes the base of the volcanic section just north of Aarja (Alabaster et al. 1982). The Bayda deposit, about 2 km northeast of Aarja, is a deeper massive sulphide located within the basal Geotimes V1 lavas or even down section in the sheeted dykes. The Lasail, Aarja and Bayda massive sulphides are interpreted as hydrothermal epithermal or mesothermal systems above hot plagiogranite intrusions with mineralised vein and stock network associated with late faults.

Fig. 8.22 Landsat photograph of the Wadi Jizzi area, northern Oman. White Cenozoic limestones of Jebel Hafit (left) and the frontal folds contrast with the black peridotites of the Mantle sequence and slightly lighter gabbros and basalts of the crustal sequence of the Semail Ophiolite. The Sumeini Window is the prominent embayment at the bottom. The Wadi Jizzi ophiolite GeoPark site lies between Wadi Jizzi in the north and Wadi Ahin in the south

Fig. 8.23 Geotimes (V1) pillow lavas, Wadi Jizzi

Fig. 8.24 Sheeted dykes at Wadi Fizh

Fig. 8.25 Wadi Jizzi ophiolite Moho dividing peridotites of the Mantle sequence (left) from gabbros of the lower crust (right)

References

Alabaster T, Pearce J, Malpas J (1982) The volcanic stratigraphy and petrogenesis of the Oman Ophiolite complex. Contrib Mineral Petrol 81(3):168–183

Cooper DJW, Searle MP, Ali MY (2012) Structural evolution of Jabal Qumayrah: a salt-intruded culmination in the northern Oman mountains. GeoArabia 17:121–150

Cooper DJW, Ali MY, Searle MP (2018) Origin and implications of a thrust-bound gypsiferous unit along the western edge of Jabal Sumeini, northern Oman mountains. J Asian Earth Sci 154:101–124

Cowan RJ, Searle MP, Waters DJ (2014) Structure of the metamorphic sole to the Oman Ophiolite, Sumeini Window and Wadi Tayyin: implications for ophiolite obduction processes. In: Rollinson HR, Searle MP, Abbasi IA, Al-Lazki A, Al-Kindy MH (eds) Tectonic evolution of the Oman Mountains. Geological Society, London, Special Publication vol 392, pp 155–176

Glennie KW, Boeuf MG, Hughes-Clarke MHW, Moody-Stuart M, Pilaar WF, Reinhardt BM (1973) Late Cretaceous nappes in the Oman Mountains and their geologic evolution. Bull Am Assoc Pet Geol 57:5–27

Glennie KW, Boeuf MG, Hughes-Clarke MHW, Moody-Stuart M, Pilaar WF, Reinhardt BM (1974) Geology of the Oman Mountains. Verhandelingen Koninklijk Nederlands geologisch mijnbouwkundidg Genootschap no. 31, 423 p

Kaneko R, Adachi Y., Miyashita S (2014) Origin of large wehrlitic intrusions from the Wadi Barghah to Salahi area in the northern Oman ophiolite. In: Rollinson HR, Searle MP, Abbasi IA, Al-Lazki A, Al-Kindy MH (eds) Tectonic evolution of the Oman Mountains. Geological Society, London, Special Publication vol 392, pp 213–228

Lippard SJ, Shelton AW, Gass IG (1986). The ophiolite of Northern Oman. Blackwell Scientific Publications, Oxford, 178 p

MacLeod C, Lissenberg J, Bibby LE (2013) 'Moist MORB' axial magmatism in the Oman ophiolite: the evidence against a mid-ocean ridge origin. Geology 41:459–462

Pearce JA, Alabaster T, Shelton AW, Searle MP (1981) The Oman ophiolite as a Cretaceous arc-basin complex: evidence and implications. Philos Trans R Soc Lond A300:299–317

Rioux M, Garber J, Bauer A, Bowring S, Searle MP, Keleman P, Hacker B (2016) Synchronous formation of the metamorphic sole and igneous crust of the Semail ophiolite: new constraints on the tectonic evolution during ophiolite formation from high-precision U-Pb zircon geochronology. Earth Planet. Sci. Lett. 451:185–195

Searle MP (1984) Alkaline peridotite, pyroxenite and gabbroic intrusions in the Oman Mountains. Can J Earth Sci 21:396–406

Searle MP, Ali MY (2009) Structural and tectonic evolution of the Jabal Sumeini – Al Ain – Buraimi region, northern Oman and eastern United Arab Emirates. GeoArabia 14:115–142

Searle MP, Malpas J (1980) The structure and metamorphism of rocks beneath the Semail Ophiolite of Oman and their significance in ophiolite obduction. Transactions Royal Society, Edinburgh. Earth Sci 71:247–262

Searle MP, Malpas J (1982) Petrochemistry and origin of sub-ophiolite metamorphic and related rocks in the Oman Mountains. J Geol Soc 139:235–248.

Searle MP, Lippard SJ, Smewing JD, Rex DC (1980) Volcanic rocks beneath the Semail Ophiolite in the northern Oman Mountains and their tectonic significance in the Mesozoic evolution of Tethys. J Geol Soc Lond 137:589–604

Skelton PW, Nolan SC, Scott RW (1990) The Maastrichtian transgression onto the northwestern flank of the proto-Oman Mountains: sequences of rudist-bearing beach to open shelf facies. In: Robertson AFH, Searle MP, Ries AC (eds) The geology and tectonics of the Oman region. Geological Society, London, Special Publication no. 49, pp 521–548

Smewing JD (1981) Mixing characteristics and compositional differences in mantle-derived melts beneath spreading axes: evidence from cyclically layered rocks in the Ophiolite of North Oman. J Geophys Res 86:80B1250

Umino S, Miyashita S, Hotta F, Adachi Y (2003) Along-strike variation of the sheeted dike complex in the Oman Ophiolite: insights into subaxial ridge segment structures and the magma plumbing system. Geochem Geophys Geosyst 4(9):1–34

Watts KF (1990) Mesozoic carbonate slope facies marking the Arabian platform margin in Oman: depositional history, morphology and paleogeography. In: Robertson AFH, Searle MP, Ries AC (eds) The geology and tectonics of the Oman region. Geological Society, London, Special Publication no. 49, pp 161–188

Watts KF, Garrison RE (1986) Sumeini Group, Oman—Evolution of a Mesozoic carbonate slope on a south Tethyan continental margin. Sed Geol 48:107–168

Ibri–Wadi Hawasina–Rustaq

9

The Hawasina Window is a large anticline that forms the northwestern extension of the Jebel Akhdar anticline (Fig. 9.1). The shelf carbonates of Jebel Akhdar form a giant, slightly asymmetric anticline with the fold axis plunging gently WNW beneath the Hawasina Window. The overlying Semail ophiolite has been folded around the anticline and crops out around the window on all sides. The Hawasina Window exposes all allochthonous units beneath the ophiolite and above the shelf carbonates. Haybi complex rocks around the margin consist of metamorphic sole, Oman Exotic limestones, Haybi volcanics and deep-water sedimentary rocks. Two of the largest and most impressive Exotics include Jebel Misht and Jebel Kawr, along the southwest flank of Jebel Akhdar. The Hamrat Duru Group rocks exposed around Jebel Milh show three large-scale fold-nappes, all backfolded, facing to the north. The structurally deepest parts of the Hawasina Window show shelf slope-margin facies Sumeini Group limestones intensely folded and thrust at Jebel Rais, Jebel Rastun and Jebel Mawq. This chapter includes the geological sites around the Hawasina Window and around the northern and southwestern margins of the Jebel Akhdar massif.

9.1 Jebel Qumayrah

Jebel Qumayrah is a structural culmination of Sumeini Group shelf margin carbonates of Triassic to Lower Cretaceous age that has folded all overlying thrust sheets (Hawasina, Haybi complexes and Semail ophiolite) over the top (Fig. 9.2). Most of the mountain massif shows two large-scale, west-vergent fold-thrust sheets–Jebel Fayyad in the west and Jebel al Huwar in the east. The rocks are

© Springer Nature Switzerland AG 2019
M. Searle, *Geology of the Oman Mountains, Eastern Arabia*,
GeoGuide, https://doi.org/10.1007/978-3-030-18453-7_9

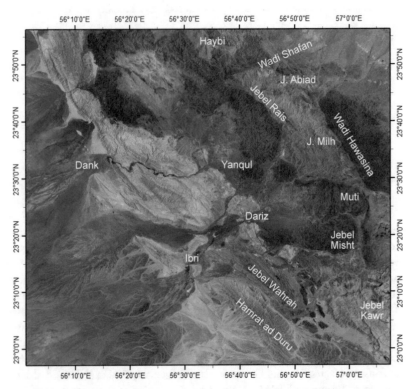

Fig. 9.1 Landsat photo of the Ibri–Wadi Hawasina region showing the key locations

Triassic to Lower Cretaceous shelf slope margin facies deposited outboard of the shelf carbonates and more proximal than the overlying Hamrat Duru Group rocks. About ten small stringers of salt have recently been discovered along Wadi Lisail and Wadi Sumer in the central part of the jebel (Figs. 9.3 and 9.4). The salt intrusion is preserved as a gypsum-anhydrite matrix enclosing blocks of Sumeini carbonates (Cooper et al. 2012). Gravity profiles suggest that the salt has intruded along a narrow fault and is almost certainly Neoproterozoic in age making it a northern extension of the Fahud salt basin (Cooper et al. 2013). The importance of the Jebel Qumayrah discovery is that the Fahud salt basin must extend north beneath the Semail ophiolite thrust sheet.

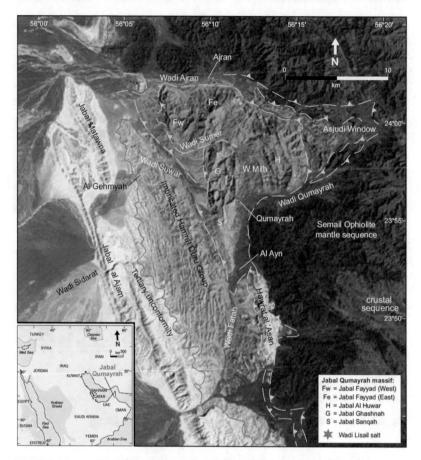

Fig. 9.2 Satellite photo of the Asjudi Window, Jebel Qumayrah region showing main structural features and the location of the salt intrusion in Wadi Lisail, after Cooper et al. (2012)

9.2 Asjudi Window

The Asjudi Window is another culmination where the Semail ophiolite has been domed and folded above a basal detachment, and eroded, exposing the complicated thrust structures immediately beneath the Semail thrust (Fig. 9.5). The Haybi

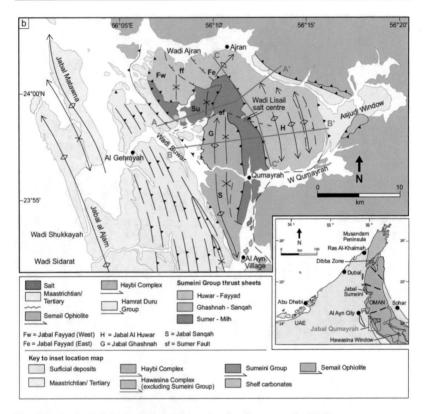

Fig. 9.3 Geological map of Jebel Qumayrah, after Cooper et al. (2013)

complex rocks (alkali volcanics, Triassic exotic limestones, cherts, metamorphic
sole amphibolites and greenschists) have been broken up into an ophiolitic mélange immediately beneath the Semail thrust (Searle and Malpas 1980, 1982; Searle
1985). This mélange is the ductile detachment horizon upon which the entire
ophiolite was emplaced onto the Arabian continental margin. Structurally beneath
the ophiolitic mélange, intact thrust sheets of distal and proximal Hawasina
complex sediments have been imbricated during emplacement. Late-stage culmination and rising of the Jebel Qumayrah Sumeini Group and shelf carbonates have
folded the thin-skinned thrust sheets above (Fig. 9.6). West of Jebel Qumayrah
imbricated sedimentary rocks of the Hamrat Duru Group are well exposed along
the foreland fold-thrust belt (Cooper 1988, 1990). These imbricated Hawasina

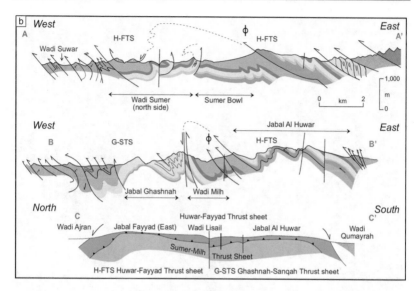

Fig. 9.4 Cross sections and a lateral section across the Jebel Qumayrah area, after Cooper et al. (2013). Lines of section are marked on map Fig. 9.3

Fig. 9.5 Aerial photograph of the Jebel Qumayrah–Asjudi Window region view towards southwest. Low brown hills of the Semail ophiolite wrap around the bedded Sumeini Group limestones of Jebel Qumayrah with white limestones of the Paleocene-Eocene in distance

Fig. 9.6 Box fold style of folding of the Sumeini Group limestones on Jebel Ghashnah–Jebel Fayyad from the Asjudi Window

complex sediments are overlain unconformably by the Maastrichtian Qahlah and Simsima Formation conglomerates and limestones which are also folded showing asymmetric west-verging folding along the frontal jebels (Figs. 9.7 and 9.8).

Access to the Asjudi Window is either along Wadi Qumayrah in the south or along Wadi Ajran, north of Jebel Qumayrah. Here, Jebel Ajran forms an isolated Oman Exotic mountain with alkali volcanic pillow lavas, red radiolarian cherts outcropping beneath Upper Triassic coral limestones (Figs. 9.9 and 9.10). An alkaline ultramafic sill has intruded the Haybi volcanics, similar to the alkali sills in Jebel Ghawil and the Hawasina Window. These are the deep roots of an off-axis alkali volcano in the Tethyan ocean. The sill comprises jacupirangite, a high-Ti rock composed of titanaugite, kaersutite amphibole and biotite, biotite wehrlite and alkali gabbro. These rocks are only found with the Haybi complex thrust sheet immediately beneath the ophiolite. The geochemistry of the Haybi lavas shows that they are highly enriched in Ti, Zr, Hf, P, Ce and Nb, the 'incompatible elements' that show they are not related to the ophiolite at all.

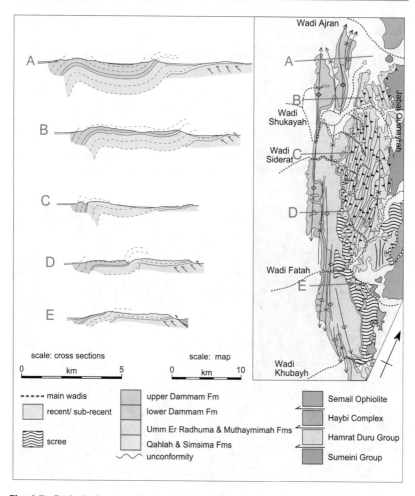

Fig. 9.7 Geological map and cross-sections across the Cenozoic jebels along the foreland west of Jebel Qumayrah

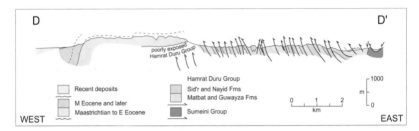

Fig. 9.8 Cross-section across the imbricate thrust slices of the Hawasina complex west of Jebel Qumayrah and the unconformably overlying Cenozoic limestones

Fig. 9.9 Jebel Ajran Exotic north of Jebel Qumayrah showing white Triassic limestones overlying dark coloured pillow lavas of the Triassic Haybi volcanics and an alkaline peridotite–gabbro sill

Fig. 9.10 Jebel Ajran Exotic limestone overlying Haybi volcanic rocks including ankaramites, alkali basalt and nephelinite

9.3   Hawasina Window and Jebel Rais

The Hawasina Window is completely surrounded by mantle sequence peridotites of the Semail ophiolite. It forms a large anticlinal culmination folded around the Jebel al-Akhdar anticline axis which extends NW from the nose of Jebel Akhdar (Fig. 9.11). The Hawasina Window exposes several large deep-level culminations of Sumeini Group slope facies carbonates (e.g. Jebels Rais, Mawq, Rastun), spectacular recumbent fold-nappes in the Hamrat Duru Group rocks of the prox-imal Hamrat Duru Group (Jebel Milh), and an extensive Haybi complex thrust sheet, sandwiched beneath the ophiolite above the Hawasina thrust sheets beneath. A cross-section of the Hawasina Window shows a series of recumbent backfolds, verging to the NE with an inverted NE limb between Wadi Hawasina and Wadi Shafan (Fig. 9.12). This NE-verging folding is thought have been caused by a steep ramp in the shelf margin during the thrust emplacement process (Searle and Cooper 1986).

The Haybi complex rocks include large Upper Triassic exotic limestones (e.g. Jebel Abiad), and associated alkaline volcanic rocks, pillow lavas with pyroxene

Fig. 9.11 Landsat photo of the Hawasina Window surrounded by dark mantle sequence harzburgites of the Semail ophiolite

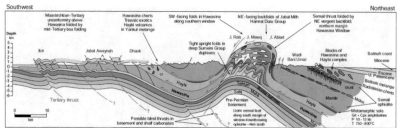

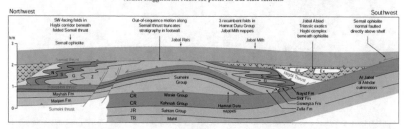

Fig. 9.12 Cross-section and a lateral section across the Hawasina Window showing geometry of the large scale thrust sheets, after Searle (2007)

Fig. 9.13 NE-directed backfolds in the Sumeini Group limestones of Jebel Mawq, viewed from Jebel Abiad Exotic above Wadi Bani Umar, northern margin of the Hawasina Window

Fig. 9.14 'Lees chain' in a deeply incised wadi cutting across Jebel Rais, Hawasina Window

(titan-augite) and plagioclase phenocrysts, intruded by rare alkali peridotite-pyroxenite sills. The Metamorphic sole rocks along the footwall of the Semail thrust including amphibolites in this region lacking garnet, meta-cherts and greenschists. The base of the ophiolite mantle sequence along Wadi Hawasina shows prominent banded olivine + orthopyroxene harzburgites with a minor amount of clinopyroxene (chrome diopside) and late hornblende. The Banded ultramafic unit is exposed above the Semail thrust and is particularly well exposed between the villages of Shakbut and Ghab in Wadi Hawasina.

Jebel Rais is one of the structurally deepest fold-thrust sheets in the middle of the Hawasina Window. It is composed of tightly folded Sumeini Group shelf edge limestones with proximal Hamrat Duru Group (Hawasina complex) sedimentary rocks folded around the top (Searle and Cooper 1986). Other Sumeini Group outcrops in the Hawasina Window include Jebel Mawq and Jebel Rastun, both of which show NE-facing tight folds and thrusts structurally beneath the Hawasina rocks (Fig. 9.13). Jebel Rais is split by a deep gorge that provides access across the range to the southwestern margin of the Hawasina Window and the base of the ophiolite (Cooper et al. 2012). An iron chain, ('Lees' chain'), has been fixed up some large boulders that block the wadi in order to gain access to the upper gorge, reputedly by G.M. Lees and K.W. Grey during their epic geological survey of the mountains for D'Arcy Exploration Company in 1925–6 (Lees 1928). The chain actually pre-dated Lees' visit as local villagers used this as a route to access the remote western side of the Hawasina Window (Fig. 9.14).

9.4 Jebel Milh Fold-Nappes

In the middle of the Hawasina Window, three very large-scale fold-nappes are beautifully exposed on the Jebel Milh massif (Searle and Cooper 1986). These tight folds affect the Hamrat Duru Group, the proximal slope facies of the Hawasina complex. The Hamrat Duru Group includes the Triassic Al Jil Formation, the Upper Triassic-Jurassic Matbat and Guwayza Formations, the Early Cretaceous Sidr cherts and the Albian-Aptian Nayid Formation limestones (Figs. 9.15 and 9.16). All these recumbent folds unusually are facing to the NE, opposite to the main emplacement direction of thrust sheets (Fig. 9.17). These NE-facing fold nappes are thought to be the result of a large promontory in the Arabian continental margin protruding northwards into Tethys. Jebel Milh can be climbed from the Wadi Hawasina in the north, Wadi ad Dil in the west and Wadi Harim in the east. The summit ridge of Jebel Milh gives an excellent panoramic view of the Hawasina Window in every direction.

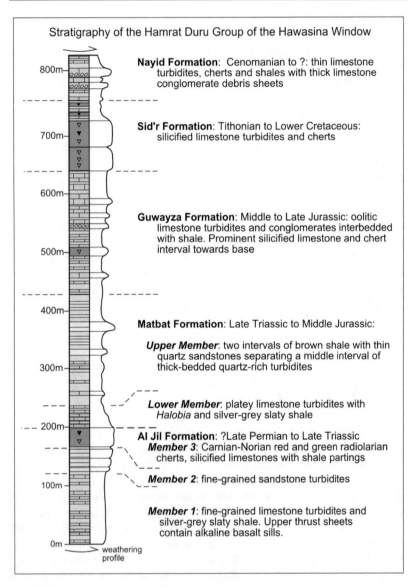

Stratigraphy of the Hamrat Duru Group of the Hawasina Window

Nayid Formation: Cenomanian to ?: thin limestone turbidites, cherts and shales with thick limestone conglomerate debris sheets

Sid'r Formation: Tithonian to Lower Cretaceous: silicified limestone turbidites and cherts

Guwayza Formation: Middle to Late Jurassic: oolitic limestone turbidites and conglomerates interbedded with shale. Prominent silicified limestone and chert interval towards base

Matbat Formation: Late Triassic to Middle Jurassic:

Upper Member: two intervals of brown shale with thin quartz sandstones separating a middle interval of thick-bedded quartz-rich turbidites

Lower Member: platey limestone turbidites with *Halobia* and silver-grey slaty shale

Al Jil Formation: ?Late Permian to Late Triassic
Member 3: Carnian-Norian red and green radiolarian cherts, silicified limestones with shale partings

Member 2: fine-grained sandstone turbidites

Member 1: fine-grained limestone turbidites and silver-grey slaty shale. Upper thrust sheets contain alkaline basalt sills.

weathering profile

Fig. 9.15 Stratigraphy of the Hamrat Duru Group in the Hawasina Window, after Cooper et al. (2014)

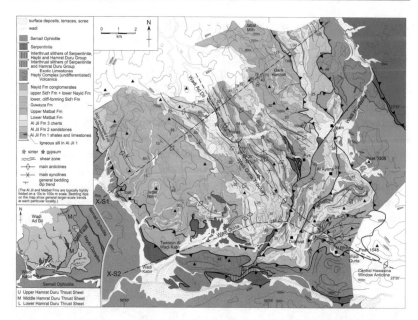

Fig. 9.16 Geological map of the southern part of the Hawasina Window, after Cooper et al. (2014)

Fig. 9.17 Recumbent NE-facing fold-nappes in the Hamrat Duru Group sediments, Hawasina Window, Jebel Milh, view west from Wadi Hawasina

9.5 Wadi Hawasina Base of Ophiolite

Along the NE margin of the Hawasina Window the base of the ophiolite is exposed along Wadi Hawasina near the villages of Budayah, Shakbut and Harmali (Fig. 9.18). The base of the ophiolite is marked by the Banded ultramafic unit, a prominent horizon of ductiley deformed alternating layers of dark brown harzburgite (olivine + orthopyroxene) and pale brown dunite (olivine). Uncommon lherzolites have been described in the lowermost ophiolite mantle sequence with up to 10% clinopyroxene (diopside) and secondary hornblende as well as olivine and enstatite. Discontinuous exposures of the metamorphic sole are exposed beneath the Semail thrust with the best exposure at the junction of Wadi Hawasina and Wadi ad Dil. Here, amphibolites with clinopyroxene pass downwards structurally into epidote amphibolites and greenschist facies cherts, marbles

Fig. 9.18 Profile across the base of the Semail ophiolite mantle sequence and Banded ultramafic unit (right), metamorphic sole and Haybi complex including the white coloured Exotic limestone block, Wadi ad Dil junction with Wadi Hawasina

Fig. 9.19 Tight to isoclinal folds and chevron folds in cherts of the Zulla formation, Wadi ad Dil, Hawasina Window

and rare meta-volcanic rocks. The whole amphibolite section has well-developed ductile mylonite fabrics. At the junction of Wadi Hawasina with Wadi al Dil, the Haybi complex includes a small Upper Triassic Exotic limestone, Jebel Harim, surrounded by distal red cherts and shales. The Haybi complex has been thrust above distal Hawasina complex cherts (Zulla or Al Jil Formation). Along Wadi ad Dil spectacular outcrops of distal cherts show tight, isoclinal to chevron folds (Fig. 9.19). All the rocks in the Hawasina Window have been folded around the NW plunging fold axis of the Jebel Akhdar anticline (Figs. 9.20 and 9.21). Two cross-sections and a lateral section across the Hawasina Window are shown in Fig. 9.22 showing the major structures and thrust sheets.

Fig. 9.20 Northwest plunging nose of the Jebel Akhdar anticline axis above Murri village, southeastern part of the Hawasina Window (photo courtesy of Janos Urai)

Fig. 9.21 Jebel Akhdar anticline plunging WNW beneath the Hawasina Window, Murri village

9.6   Jebel Misht Exotic

Jebel Misht (2090 m) is the most impressive of all the 'Oman Exotic' limestones. Jebel Misht is composed of well-bedded Upper Triassic reef limestone approximately 900 m thick, above thin discontinuous exposures of Triassic alkali basalts (Haybi volcanics). These reefal limestones contain abundant corals and large bivalves, *Megalodon* sp. and have been interpreted as representing either a rifted part of the Jebel Akhdar platform, or as an oceanic guyot or seamount above an alkali basaltic ocean island (Searle and Graham 1982). The limestones have been affected by metamorphism from the metamorphic sole and are recrystallised in places as marble. The Haybi thrust sheet lies structurally above the distal Hawasina complex thrusts sheets, comprised mainly of thin-bedded red radiolarian cherts (Fig. 9.23).

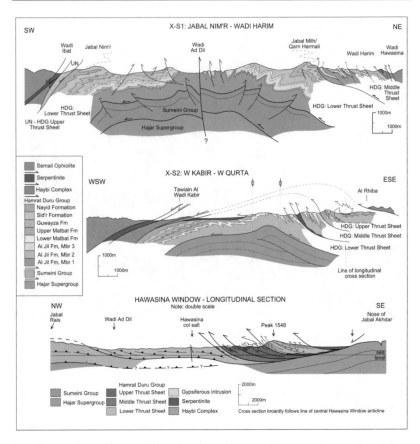

Fig. 9.22 Cross sections and a lateral section across the Hawasina Window, after Cooper et al. (2014)

The spectacular cliffs of the south face are 1000 m high and near vertical, providing the best rock-climbing crag in all the Oman-UAE mountains (Figs. 9.24 and 9.25). Several impressive climbing routes now adorn the south and southeast faces, with the 'Nose' of Jebel Misht being the most exciting (Fig. 9.26). Shorter but equally exciting rock climbs can be made up the western end of Jebel Misht (Fig. 9.27). Climbs can be done in one day with an early start from Wadi al Ayn, and a descent down the north flank. The northern flank is a gentle 30° dip-slope extending to the base of the Semail ophiolite and can be used as a descent route.

Fig. 9.23 South face of Jebel Misht Exotic showing 1000 m high cliffs of Upper Triassic limestone overlying thin alkali basalts (Haybi volcaics) and structurally overlying thin bedded red radiolarian cherts of the distal Hawasina complex

9.7 Wadi Damm, Sint Ophiolite Klippe

Wadi Damm drains west from Jebel al-Akhdar–Jebel Shams, and runs between the two large exotic blocks of Jebel Misht to the north and Jebel Kawr to the south. It is accessed from the main Nizwa to Ibri road along the Wadi al Ayn. The prehistoric 'beehive' tombs of Bat are built on red radiolarian cherts and have an impressive backdrop with the vertical cliffs of Jebel Misht behind. These tombs date from the third millennium BC and, together with the better-preserved tombs of Al Ayn and Al Khutm, are a UNESCO World heritage site of great archaeological importance. An ancient unusual square-shaped Omani fort in the village is surrounded by date gardens (Fig. 9.28). Wadi Damm is a scenically beautiful wadi with several pools fed from springs in the limestone. One particular pool is turquoise coloured with a small waterfall, flanked by dripping ferns and green moss. It is possible to trek to Jebel Shams from this wadi past the settlement of Misfat al Khaqatur. South of Wadi Damm is the Sint syncline, a shallow bowl of ophiolite

Fig. 9.24 Main cliffs of the Triassic exotic limestones on the south Face of Jebel Misht

Fig. 9.25 Final two pitches to the summit ridge of Jebel Misht south face Jebel Misht

Fig. 9.26 'Nose' route on Jebel Misht south face

peridotite surrounded by the high Exotic limestone peaks of Jebel Kawr. The Sint ophiolite is a remnant klippe of the main Semail ophiolite that has been folded around and above the Jebel Akhdar anticline. Like the Bahla- Nizwa ophiolite, the

Fig. 9.27 Jebel Misht exotic limestones from west

Fig. 9.28 Unusual square fort at Wadi Sint, Jebel Kawr Exotic in background

Sint syncline is a thin thrust sheet of ophiolite preserved along the southern flank of Jebel Akhdar. The structures clearly show that the ophiolite and underlying thrust sheets have been folded after emplacement onto the continental margin.

9.8   Jebel Kawr Exotic

Jebel Kawr, one of the largest Exotic limestones in Oman is a high massif flanked by steep cliffs all around to the south of Jebel Misht (Fig. 9.29). A deeply incised wadi cuts into the western ramparts of Jebel Kawr and a narrow track leads to the hidden village of Nadan. Jebel Kawr is geologically similar to Jebel Misht with about 1000 m of Upper Triassic limestone above an alkali volcanic substrate (Haybi volcanics). Above the limestones about 20 m thickness of Jurassic cherts and shales (Nadan Formation) overlies the limestone. This suggests drowning of the Triassic oceanic island or guyot at the end of the Triassic. Jebel Kawr lies structurally above the Hawasina complex rocks around Wadi Al Ayn and structurally beneath the peridotites of the Sint syncline. A series of other isolated Oman Exotic limestones

Fig. 9.29 1500 high limestone cliffs on the south face of Jebel Kawr exotic above Al Hayl village, Wadi al Ayn

overlies the Cretaceous shelf carbonates all the way over the Jebel Akhdar anticline. Jebels Kawr, Misht and Misfah are all within the Haybi complex beneath the ophiolite and above the shelf carbonates (Searle and Graham 1982).

9.9 Wadi Abiad Ophiolite Section

Wadi Abiad ('white wadi' in Arabic) shows an excellent profile across the ophiolite sequence and is the closest of the classic ophiolite profiles to Muscat. Road access is from the Batinah coast in the north to the village of Al Khatum and thence by dirt track along the wadi, or from the south from the Awabi–Nakhl road to the small village of Subakhah. These roads do not connect across the harzburgite mountains. The ophiolite dips to the north, so approaching from the Batinah coastal plain, progressively deeper rocks are exposed to the south. Scattered outcrops of Batinah mélange, pillow lavas and sheeted dykes are present, but the lower gabbros make up the cliffs bounding the wadi. Depending on the water level in the wadi it is possible to drive some 4 km along the wadi until the track ends; from here it is 1–2 km walking to get to the site of the Moho.

Wadi Abiad provides one of the best sections across the Moho in Oman. The mantle section is composed entirely of harzburgite (olivine + enstatite) and dunite (olivine). Layering increases upwards towards the Moho, and it has been suggested that the Moho acted as a ductile shear zone because of the different rheological properties of the ultramafic mantle rocks and the plagioclase-bearing crustal rocks. Small chromite bodies occur along the upper part of the mantle section. The Moho is in places a sharp contact, and elsewhere is a zone of interbanded peridotites and gabbros. The gabbros have cumulus textures with olivine and pyroxene first to crystallise, forming the cumulus minerals, and plagioclase forming the inter-cumulus. These textures are classic examples of gravity settling along the base of a magma chamber. Small dark-coloured wehrlite (olivine + clinopyroxene) intrusions cut the layered gabbros and are thought to be related to arc-related volcanics of the V2 (Lasail) volcanic rocks.

9.10 Wadi Hamaliyah Ophiolite Section

Wadi Hamaliyah is located about 150 km west of Muscat and is one of many wadis cutting across the Rustaq ophiolite block. This profile shows a continuous section from the mantle sequence and Moho Transition zone up to 450 m thick, with a 3 km thick section through lower crust gabbros (Rollinson 2009, 2014).

Approaching from the north, the first outcrops are blocks of Hawasina-type sediments, part of the Batinah mélange that structurally overlies the ophiolite. The volcanic sequence with inter-bedded umbers is not well exposed in this area but the sheeted dykes appear near the village of Huwayl with small plugs of trondhjemites or plagiogranites, comprised almost entirely of plagioclase with minor amounts of quartz, hornblende and chlorite, and thought to represent the final fractionated melts from the ophiolite suite. Homogeneous gabbros (clinopyroxene + plagioclase), or gabbro-norite (with minor orthopyroxene) and leucogabbros (plagioclase + olivine + clinopyroxene) make up the upper part of the gabbro section. The lower part is comprised of layered gabbros and olivine-rich troctolites (olivine + plagioclase) with small, 10–20 cm thick, brown sulphide-rich (pyrite, pyrrhotite, chalcopyrite) layers. Minor amounts of hornblende-bearing plagiogranite form magmatic breccias intrusive into gabbros. The Moho is exposed near the tiny village of Hayl Al-Ghafar and is relative abrupt with a limited transition zone. The uppermost mantle rocks consist of dunite (only olivine), pyroxenite and wehrlite (olivine + clinopyroxene). The wehrlite and dunite sequence has been interpreted as a 300 m thick intrusion into the base of the oceanic crust. Beneath the Moho the true mantle rocks comprises harzburgite (olivine + orthopyroxene) and dunite representing the depleted mantle section. These ultramafic mantle rocks continue to the west across the mountain range.

References

Cooper DJW (1988) Structure and sequence of thrusting in deep-water sediments during ophiolite emplacement in the south-central Oman Mountains. J Struct Geol 10:473–485

Cooper DJW (1990) Sedimentary evolution and palaeogeographical reconstruction of the Mesozoic continental rise in Oman: evidence from the Hamrat Duru Group. In: Robertson AFH, Searle MP, Ries AC (eds) The geology and tectonics of the Oman region. Geological Society, London, Special Publication no. 49, pp 161–188

Cooper DJW, Searle MP, Ali MY (2012) Structural evolution of Jebel Qumayrah: a salt-intruded culmination in the northern Oman Mountains. GeoArabia 17:121–150

Cooper DJW, Ali MY, Searle MP, Al-Lazki AI (2013) Salt intrusions in Jabal Qumayrah, northern Oman Mountains: implications from structural and gravity investigations. GeoArabia 18:141–176

Cooper DJW, Ali MY, Searle MP (2014) Structure of the northern Oman mountains from the Semail Ophiolite to the foreland basin. In: Rollinson H, Searle MP, Abbasi I, Al-Lazki A, Al-Kindy M (eds) Evolution of the Oman mountains. Geological Society, London, Special Publication no 392, pp 129–153

Lees GM (1928) The physical geography of South-Eastern Arabia. Geogr J 71:441–466

Rollinson H (2009) New models for the genesis of plagiogranites in the Oman ophiolite. Lithos 112:603–614

Rollinson H (2014) A (virtual) field excursion through the Oman ophiolite. Geol Today 30 (3):110–118

Searle MP (1985) Sequence of thrusting and origin of culminations in the northern and central Oman Mountains. J Struct Geol 7:129–143

Searle MP (2007) Structural geometry, style and timing of defrmation in the Hawasina window, Al Jabal al-Akhdar and Saih Hatat culminations, Oman mountains.GeoArabia 12:93–124

Searle MP, Cooper DJW (1986) Structure of the Hawasina window culmination, central Oman mountains. Trans R Soc Edinburgh Earth Sci 77:143–156

Searle MP, Malpas J (1980) The structure and metamorphism of rocks beneath the Semail ophiolite of Oman and their significance in ophiolite obduction. Trans R Soc Edinb Earth Sci 71:247–262

Searle MP, Malpas J (1982) Petrochemistry and origin of sub-ophiolite metamorphic and related rocks in the Oman Mountains. J Geol Soc Lond 139:235–248

Searle MP, Graham GM (1982) "Oman Exotics"—oceanic carbonate build-ups associated with the early stages of continental rifting. Geology 10:43–49

Jebel al-Akhdar

10

The Jebel Akhdar (al Jebel al-Akhdar, 'Green Mountain' in Arabic) massif forms the highest part of the Oman Mountains, culminating in Jebel Shams ('Mountain of the Sun' in Arabic) at 3009 m altitude (Fig. 10.1). It is a scenically beautiful area with rugged steep cliffs along the northern side, deeply incised wadis with perennial flowing streams, narrow canyons with blue pools and waterfalls, and beautiful little villages with green terraced fields clinging to the side of steep wadis. The Saiq plateau in the eastern part is a high and relatively flat area with steep-sided wadis draining off in all directions. During rainy periods waterfalls suddenly appear and lower parts are particularly prone to flash flooding.

The Jebel Akhdar massif is a very large scale 60 km wavelength anticline plunging to the northwest beneath the Hawasina Window. The fold axis curves from WNW-ESE (Jebel Shams) to NNE-SSW (Jebel Nakhl) and back to E-W aligned towards Semail (Fig. 10.2). This is a fold interference pattern caused by a dominant NE-SW compressive stress, accompanied by a subsidiary WNW-ESE compression forming the Jebel Nakhl anticline trend. The shelf carbonates are remarkably conformable showing only a few minor folds due to intra-formational thrusting during folding (Fig. 10.3). The Jebel Akhdar anticline has folded the Permian to Late Cretaceous shelf carbonates and the underlying pre-Permian basement rocks as well as all the overlying Late Cretaceous thrust sheets including the Semail Ophiolite, Haybi and Hawasina complexes (Fig. 10.4). The upper structural boundaries around the flanks have been reactivated as normal faults during the uplift of the structure. Balanced cross-sections suggest that a major detachment or blind thrust probably occurs beneath the Jebel Akhdar anticline at depth, ramping up from deep levels in the north towards the south, to underlay the outer folds seen in the foothill jebels of the Salakh arch in the far south of the mountain range.

© Springer Nature Switzerland AG 2019
M. Searle, *Geology of the Oman Mountains, Eastern Arabia*,
GeoGuide, https://doi.org/10.1007/978-3-030-18453-7_10

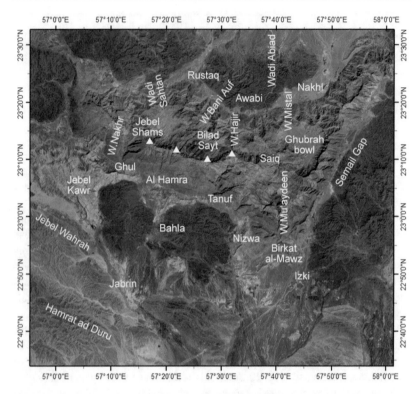

Fig. 10.1 Landsat photograph of the Jebel al-Akhdar massif showing main localities and towns

Some major north-south aligned wadis (Sahtan, Bani Awf, Bani Kharus, Mistal) cut through the entire massif at right-angles to the strike and provide ideal cross-sections to study the Permian to Mesozoic stratigraphy of the shelf carbonates as well as the pre-Permian basement in the core of the fold. Intense folding, thrusting and cleavage development in pre-Permian rocks reveal that a Late Paleozoic (Devonian-Carboniferous) orogeny occurred in Oman; these structures are abruptly truncated beneath the Middle Permian unconformity. The best wadi sections through the pre-Permian basement rocks occur in Wadis Mistal, Bani Kharus and Bani Awf, where evidence for two Neoproterozoic ('Snowball Earth') glacial events (Ghaba and Fiq glacials) is seen.

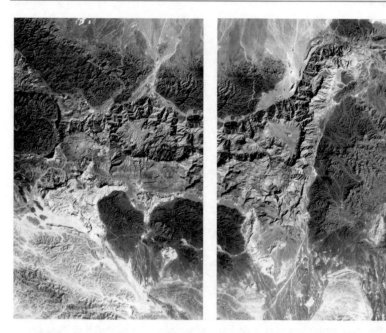

Fig. 10.2 Landsat photograph of the Central Oman Mountains showing the Jebel al-Akhdar window surrounded by the overthrust Semail ophiolite. The core of the Jebel al-Akhdar anticline is the Pre-Permian basement. South of the Bahla ophiolite block are the imbricated Hawasina complex sediments with large Oman Exotic limestone blocks (Jebel Kawr and Jebel Misht in the west. The NNE-SSW trending Semail Gap separates Jebel Nakhl and Jebal al-Akhdar to the west from the Ibra ophiolite block to the east. Photo courtesy of Petroleum Development (Oman) Ltd (Hughes-Clarke 1990)

Jebel Akhdar is the most important geological site for exposures of the Permian–Cretaceous shelf carbonate sequence in the Arabian Peninsula. The main oil reservoir horizons (Middle Cretaceous Shuaiba and Natih Formations), so important for oil and gas reserves in the foreland, are all completely exposed along numerous wadi profiles. Neoproterozoic glaciation events resulted in deposition of glacial tills and diamictites with groove marks and dropstones all of which are well exposed in the Mistal Formation in Wadi Mistal and the Fiq Formation in Wadi Sahtan, Wadi Hajir and Wadi Bani Kharus.

Fig. 10.3 Sahtan bowl view north from Jebel Shams. The Mid-Permian unconformity with Saiq Formation above basement is clear around the mountains west of Wadi Sahtan

The Jebel Akhdar area is also the most scenically attractive in all Arabia with small villages and terraced fields nestling beneath steep cliffs. Many small villages are scattered throughout the Sahtan and Ghubrah 'bowls' and the steep cliffs above are also a rock climbers and trekkers paradise. Snake Gorge in Wadi Bani Auf is one of the best canyoning expeditions anywhere in Arabia with approximately 3 km of deeply incised wadi with numerous pools and waterfalls. At least four climbing routes ascend the steep cliffs on the north face of Jebel Akhdar to the summit ridge at Jebel Shams. It has been proposed that the high ridge of Jebel Akhdar from the Jebel Shams summit extending east and west including all the steep cliffs along the north face (Figs. 10.5 and 10.6) should be designated a World Heritage Site and a National GeoPark.

Fig. 10.4 Aerial photograph of the Jebel al-Akhdar anticline with Semail ophiolite around the margins, photo courtesy of Petroleum Development (Oman) Ltd (Hughes-Clarke 1990)

10.1 Birkat al Mawz, Wadi Muaydeen

The profile along Wadi Muaydeen provides the type section through the Permian and Mesozoic shelf carbonates and has been extremely well documented since the initial surveys by PDO oil company geologists, summarized by Wilson (1969) and Glennie et al. (1973, 1974) and Hughes-Clarke (1990). Wadi Muaydeen cuts across the southern flank of Jebel Akhdar draining south from the Saiq Plateau. The town of Birkat al Mawz ('well of the banana' in Arabic) lies at the mouth of

Fig. 10.5 Aerial photograph of the north face of Jebel Shams, above Bilad Seit village and upper Wadi Bani Auf. The mid-Permian unconformity is clearly visible beneath the big limestone cliffs

Fig. 10.6 Aerial photograph of the cliffs along the eastern margin of Jebel Shams

the wadi and has a nicely restored old Omani fort (Fig. 10.7). 400 m north of the fort the road enters the lower end of the wadi. The full stratigraphic section along the Wadi Muaydeen travese is shown in Fig. 10.8. The allochthonous Hamrat Duru Group (Hawasina complex) rocks tectonically overlies the shelf carbonates along the Hawasina thrust. The Hawasina complex comprises Triassic to Cretaceous basin proximal facies sandstone and limestone turbidites (**Zulla and Guwayza Formations**) overlain by thin-bedded cherts and silicified limestones (**Sidr Formation**). The discovery of older distal Hawasina oceanic sedimentary rocks overlying younger more proximal Cretaceous shelf carbonates led to the proposal of a thrust origin for the Oman Mountains, first suggested by Glennie et al. (1973, 1974).

Continuing north up Wadi Muaydeen the Hawasina series can be seen overlying thin Aruma shales and then the top of the shelf carbonates. The Aruma Group (Turonian–Campanian) shales are the flexural foreland basin deposits into which the allochthonous thrust sheets were emplaced during the Late Cretaceous. The **Muti Formation** is at the base of the Aruma Group and is here marked by

Fig. 10.7 Old photograph (ca. 1976) of the Fort at Birkat al-Mawz, and the entrance to Wadi Mu'aydin. The ridge of Guwayza Formation is overlain by Sidr cherts (Hamrat Duru Group)

Wadi Mi'Aidin: Mesozoic Carbonate Platforms

STRATIGRAPHY - WADI MI'AIDIN

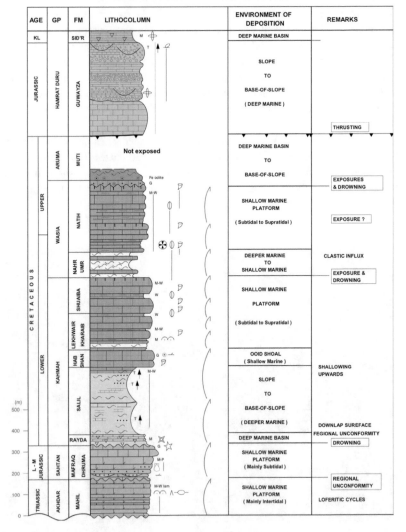

Fig. 10.8 Stratigraphic log section of the Wadi Mu'aydin (Mi'Aidin) profile across the Triassic to Cretaceous shelf carbonates and the overthrust Hamrat Duru Group (Hawasina complex), photo courtesy of Petroleum Development (Oman) Ltd (Hughes-Clarke 1990)

reworked ferruginous oolites and channels of calc-arenite and conglomerate overlain by yellow silty marls. Planktonic foraminifera indicate deep open marine conditions and sudden flooding of the sub-aerially exposed and channelised Cenomanian–Turonian Natih platform. The road to Saiq heads out of the wadi up the mountains to the northeast. The main wadi continues to the north and becomes a narrow gorge cut though the shelf carbonates of the Wasia, Kahmah and Sahtan Groups. The Triassic to Cretaceous shelf carbonate exposures along Wadi Muaydeen form some of the classic exposures of the Arabian shelf carbonate sequence and have been extensively studied. The stratigraphic logged sections along Wadi Muaydeen are direct analogues for the sequences drilled by many of the oil and gas wells in the interior. The following description goes from younger to older stratigraphic units, going from south towards the north, into the core of the Jebel Akhdar anticline.

The **Natih Formation** (Cenomanian–Lower Turonian) is approximately 280 m thick and consists of limestones, alternating with calcareous shales and marls. Van Buchem et al. (2002) and Droste (2005) describe repetitive sedimentary cycles, each consisting of a thick carbonate and then thin mixed carbonate–clastic unit. Seven cycles within the Natih Fm have been closely correlated with wells in the sub-surface. The **Nahr Umr Formation** (Late Aptian–Albian) deep to shallow marine limestones underlies the Natih Fm. This is a clastic mudstone unit widely developed over Arabia and its chief significance is as a regional top seal to the formations below. In south Oman it was accompanied by a major flooding event and rests unconformably on reservoirs as old as Ordovician. The **Shuaiba Formation** (Lower Aptian) is 87 m thick in Wadi Muaydeen and consists of rudist-bearing wackestones to packstones overlying transgressive shallow water algal platform facies (Van Buchem et al. 2002). The **Kharaib Formation** (Barremain; lowermost Aptian) underlies the Shuaiba Fm. and consists of massive rudist-bearing wackestones and packstones (52 m) overlain by nodular *Orbitolinid*-bearing wackestones and packstones (20 m) also called the *Palorbitolina* Member (P1) (Van Buchem et al. 2002; Droste 2005). These beds are followed by massive bedded, bioturbated dolomitised mudstones. The top of the Kharaib Fm. is marked by a second *Palorbitolina* bed (P2), 30 m thick, with nodular argillaceous wackestones with *Orbitolinids* and corals (Van Buchem et al. 2002; Droste 2005). The **Lekhwair Formation** (Hauterivian–Barremian) is about 155 m thick and consists of argillaceous limestones and more open marine wackestones to grainstones with small rudist biostromes. The basal part was deposited in a shallow marine, supra-tidal environment; the upper part in a more open marine setting (Droste 2005). The **Habshan Fm.** (Late Valanginian–Hauterivian) is about 44 m thick and was deposited in a very shallow marine, high-energy environment with carbonate shoals. Massive grainstones with rudists and corals grade into bioclastic oolitic grainstones with cross-bedding and containing rudists, gastropods and coral fragments (Droste 2005). In the wadi walls it displays tens of meters scale fore-seting, representing progradation of the carbonate platform towards the north. The **Salil Formation** (Berriasian–Valanginian, possibly Hauterivian in upper part; Simmons 1994) is about 330 m thick and consists of black argillaceous lime mudstones. Fine clastics and turbidites indicate deposition in a proximal basin. Bioclastic material increases upwards and there is a gradation into the overlying cliff-forming Habshan Formation with increasing bed thickness. The **Rayda Formation** (Early Berriasian–possibly earliest Valanginian; Simmons 1994) is about 130 m thick and consists of grey-weathering thin-bedded lime mudstones with conspicuous black chert layers and nodules. The fine-grained limestones contain radiolaria, sponge spicules, pelagic crinoids and tintinnids indicating earliest Cretaceous–latest Jurassic age. The Rayda cherts formed in a deep marine, bathyal environment, starved of clastic input (Droste 2005). Older descriptions refer to these beds as "*Porcellanites*". An upward increase in thicker bedded carbonates marks a gradual transition into the overlying Salil Fm.

The Jurassic **Sahtan Group** (Pleinsbachian to Oxfordian; Glennie et al. 1974) consisting of the **Mafraq and Dhruma Formations** rests with a slight angular unconformity on the Triassic Mahil Formation below. A basal conglomerate containing reworked Triassic blocks lies above a major Triassic-Jurassic uncon-formity indicating transgressive flooding of the subaerially exposed Triassic. From the base upwards mixed siliciclastics and carbonates of the Mafraq Formation with characteristic dun-brown colours change upwards to limestones. *Lithiotis* bivalves are common in several horizons. Thin muddy deeper water deposits in middle Sahtan Group mark the deepening prograding carbonate ramp. Cyclic deposition is apparent with shallowing-upward cycles. A major mid-Jurassic unconformity at the top of the Sahtan Group results in variable thicknesses of Sahtan Group. Mixed siliciclastics and carbonates change upwards to limestones. *Lithiotis* bivalves are common in several horizons. The Triassic **Akhdar Group** consists of about 600 m of massive grey dolomite (**Mahil Formation**) deposited in a shallow marine set-ting ranging from lagoonal to sub-tidal. The easily accessible part of Wadi Muaydeen ends at the small date palm gardens of Muaydeen village where the traverse continues by trekking along the boulder-strewn wadi. The Middle and Upper Permian **Saiq Formation** comprises cliff-forming well-bedded massive limestones that lie above the prominent unconformity. The dips of the beds decrease towards the north and the Saiq plateau region, toward the core of the main anticline.

10.2 Saiq (Sayq) Plateau

The Saiq plateau is a large area of high plateau between 2000 and 2400 m altitude in the middle of the Jebel Akhdar anticline with the Permian–Mesozoic carbonates dipping off to the north, east and south of the central dome. Towards the west, the Jebel Akhdar anticline becomes a tighter fold with the axis plunging NW beneath the Hawasina Window (Chap. 9). The Middle and Upper Permian Saiq Formation forms the base of the Tethyan shelf carbonate sequence and unconformably overlies all lower units, including the Precambrian Muaydeen siltstones in the Saiq plateau region. Access to the Saiq plateau is only by 4-wheel drive vehicle from Wadi Muaydeen and Birkat al Mawz. The road meanders up the steep cliffs to the ridge overlooking the Semail Gap. From here there are spectacular views towards the east over the ophiolite mountains around Semail. The road then heads west winding around the headwalls of Wadi Muaydeen to the town of Saiq. Thirty years ago Saiq comprised only a few small villages with an army base and small landing strip for STOL planes. It has expanded rapidly and now includes new housing estates in Saiq itself and a 5-star hotel resort, the Anantara resort. Shariyah was a small village of mud-brick houses and a mass of lush green terraced barley fields, famous for its limes, peach, apricot, walnut, almond, fig and pomegranate trees (Fig. 10.9). Saiq is

Fig. 10.9 Old photograph (ca. 1970) of Shariyah, Al-Ain village on the Saiq plateau showing irrigated fields of alfalfa and cherry and apricot trees in blossom

Fig. 10.10 Old photograph (ca. 1984) of Bani Habib village on the Saiq plateau

renowned for its pleasant cool climate as well as its fruit and production of rose water. The terraces and old villages are deserted now and water has to be pumped up from coast. The ancient village of Bani Habib is another deserted old Omani village tucked into a small wadi just west of Saiq (Fig. 10.10). The Saiq plateau and high mountains of Jebel Akhdar have a microclimate of their own with unique vegetation that includes stunted junipers, and wild olive trees. During harsh winter seasons, it has been known to snow on the Saiq plateau and Jebel Shams.

10.3   Wadi Kamah to Bani Habib

The stratigraphy of the Permian and Triassic shelf carbonates along Wadi Kamah is similar to that seen in Wadi Muaydeen. The trek along Wadi Kamah to Bani Habib and the Saiq plateau is a wonderful route that takes between 6 and 9 hours depending on the conditions (Dale and Hadwin 2001). It starts at Kamah village at 600 m altitude and winds its way up the limestone crags with increasingly spectacular views of the great wadi canyon (Figs. 10.11 and 11.12). Small pools and waterfalls are constantly in view and at 1450 m an isolated and deserted old mosque, Masjid Shughah mosque can be seen. The poorly defined trail winds its

Fig. 10.11 Waterfalls and pools in the upper Wadi Kamah on the trek up to Saiq

way up the dip-slope of Jebel Akhdar following the Natih Formation, until small cliff sections progressively traverse lower stratigraphic units towards the core of the Jebel Akhdar anticline. At about 1800 m the trail breaks out of the gorge as the plateau is reached. The route then approaches Saiq via the small canyon that is the location for the now-deserted village of Bani Habib.

Fig. 10.12 Wadi Kamah waterfall, southern flank of Jebel Akhdar

10.4 Wadi Tanuf

Wadi Tanuf is another spectacular gorge incised into the southern part of the Saiq plateau (Figs. 10.13 and 10.14). The village at Tanuf was the home of Sulaiman bin Himyar, the so-called 'Lord of the Green Mountain', and the center of the rebellion against Sultan Said bin Taimur of Muscat in the 1950s, but was destroyed by the Sultan's Air Force with the help of the SAS. The rebels fled up into the

Fig. 10.13 Wadi Tanuf gorge cutting into the Saiq plateau

Fig. 10.14 View south from the canyon rim, Wadi Tanuf. Note the few houses on the upper ledge

mountains but were flushed out during the subsequent jebel war. The wadi cuts through the Permian to Cretaceous shelf carbonates that are similar to the section described above for Wadi Muaydeen. Whereas it is possible to drive along Wadi Muaydeen as far as the Triassic, Wadi Tanuf requires climbing and trekking. The trail winds up the steep crags with beautiful little waterfalls and rock pools. Opening out onto the Saiq plateau at the top of a 2000 m climb, it is possible to loop around to the east and end at the old, now deserted, village of Bani Habib, tucked down in a small wadi on top of the Saiq plateau. To the west of Wadi Tanuf a new road zig-zags up onto the main Jebel Akhdar ridge at 2000 m to the tiny village of Qiyut. This road provides panoramic views along the southern dip slope of Jebel Akhdar and down to the large green oases and date gardens around Nizwa and Bahla. The Natih Formation forms much of this dip slope and numerous fossils, particularly rudists, echinoids, gastropods and bivalves can be found lying loose on the surface (Fig. 10.15).

Ghubrat Tanuf is a locality ~ 5 km east of Wadi Tanuf and a good locality to study the fault and fracture systems in the Natih Formation. Fault orientation is dominantly east-west (100°) with a secondary set oriented $\sim 130°$. The faults and

Fig. 10.15 Gastropods, bivalves and echinoid fossils from the Natih Formation on Jebel Akhdar

fractures probably act as conduits for fluids when they opened and many are now filled with calcite. Unlike similar factures reported from the oilfields SW of the mountains, this locality has been buried deeper because of the overthrust Semail ophiolite together with the underlying Haybi and Hawasina complex thrust sheets. Field geological mapping has described three sets of fractures in northern Oman: (1) related to NE-SW compressive stress during Late Cretaceous emplacement of the ophiolite and related Tethyan thrust sheets (Searle 2007), (2) one set related to WNW-ESE compressive stress as a result of either Late Cretaceous collision of Arabia with India (Filbrandt et al. 1990), or to WNW-emplacement of the Masirah ophiolite onto the south Arabian continental margin (Schreurs and Immenhauser 1999), (3) one arcuate set of mid-Cenozoic fractures related to folding of the Jebel Akhdar–Saih Hatat culmination probably initiated during the final stage of Late Cretaceous nappe emplacement. The final stage of uplift was probably Late Miocene–Pliocene in age.

10.5   Al Hooti Cave

Al Hooti cave (also called Al Fallah cave) is 4.5 km long and extends beneath the southern slopes of Jebel Akhdar. It has numerous stalactites, stalagmites and flowstone structures as well as an underground lake more than 700 m long, home to a species of cave-dwelling small white blind fish (*Garra barreimiae*). The cave has three levels, each leading to lakes. During rains the underground wadi systems of Jebel Akhdar are prone to extreme flooding. The Al Hooti cave has now been developed into a tourist attraction. It is likely that the entire Jebel Akhdar massif is riddled with similar but smaller cave systems and it is possible that some of these may inter-connect from the summit ridge area down to the base of the mountain at Al Fallah, a vertical distance of nearly 2000 m.

10.6   Wadi Nakhr Grand Canyon, Wadi Ghul

The Grand Canyon of Wadi Nakhr (Figs. 10.16, 10.17 and 10.18) is perhaps the most impressive of all the canyons of Jebel Akhdar. Successive wadi floods have carved a deep and spectacular canyon from the summit of Jebel Shams (3009 m) south to the village of Ghul. The wadi is called a variety of names including Wadi Ghul (lower part), Wadi Nakhr and Wadi Saydran (upper part). Just before the entrance to the narrow section of Wadi Ghul is an excellent exposure of the main source rock for much of North Oman's oil, the black fetid limestones of the Natih

Fig. 10.16 Jurassic–Cretaceous shelf carbonates exposed along the flanks of the Grand Canyon of Oman, Wadi Nakhr. *Note* the green date palms in small villages along the wadi floor. The Upper Triassic exotics of Jebel Kawr and Jebel Misfah are in the distance

'B' division which formed at a time of very widespread, if not global anoxia, the 'oceanic anoxic event'.

The road from Ghul village follows the west rim of the wadi canyon along the top shelf - Hawasina/Haybi complex contact, then winds up the dip slope of the Natih Formation towards the summit of Jebel Shams (3009 m). Views in all directions get better the higher up you go. Large mountain size Triassic exotic limestones of Jebel Kawr to the south, Jebel Misfah and Jebel Misht to the west are Upper Triassic carbonates sitting above a major thrust (Haybi thrust) that has emplaced Triassic rocks over the Cretaceous Natih Formation. Excellent views east across the canyon show steep cliff sections through the Cretaceous and Jurassic shelf carbonates. Sets of vertical fractures can be seen cutting the entire stratigraphic section.

Several places along the new road expose pavements in the Natih Formation where calcite filled fractures are well exposed. The *en echelon* fracture systems can be used to deduce the stress field operative during the Late Cretaceous that can be related to the flexural slip folding of the Jebel Akhdar anticline (Holland et al. 2009). The flanks of the canyon are also good locations to see vertical fault and

Fig. 10.17 Western flank of the Grand Canyon of Wadi Nakhr

fracture systems that can be related to fluid pathways from deep to shallow. These fractures have relevance to oil migration fractures in the foreland. Numerous Cretaceous fossils can be found on the surface on the canyon rim including echinoids, brachiopods and gastropods, eroded from the Natih Formation.

There are several trekking routes from Ghul village up the Jebel al-Akhdar with total ascents of up to 2300 m, and the region has rich potential for remote and hidden rock climbing routes. Possibly the most amazing trekking and climbing region is the maze of wadis leading up to the 'inaccessible pinnacle' in Wadi Saydran—a free-standing stack split by two branches of the wadi, at the top of the

Fig. 10.18 Cretaceous shelf carbonates on the rim of the Grand Canyon, Wadi Nakhr

canyon (Fig. 10.19). The 'Balcony trek' is a scenic three-hour hike that traverses
the rim of the canyon to a small deserted village with hanging terraces and a small
wadi pool (Figs. 10.20 and 10.21). Polished boulders of limestones rich in gastropod faunas can be seen in the wadi bedrock (Fig. 10.22). It is forbidden to enter
the military zone with radar station at the summit of Jebel Shams and this region

Fig. 10.19 Upper reaches of Wadi Nakhr Grand Canyon with Jebel Shams summit ridge in clouds. Note the free standing pinnacle at the junction of the upper part of the wadi

Fig. 10.20 Jurassic and Cretaceous shelf carbonates in upper Wadi Nakhr. Trekking routes from Ghul village ascend through these cliffs to the 'balcony'

Fig. 10.21 Abandoned terraced fields and small huts above overhanging cliffs beneath Jebel Shams

should be avoided. Local villagers from Ghul are famous for weaving colourful rugs made from sheep and goat wool.

10.7   Wadi Sahtan Shelf Carbonates

Wadi Sahtan cuts a deep gorge through the Mesozoic and Permian shelf carbonates from the northwest flank of Jebel Akhdar into the 'Sahtan bowl' the area of Precambrian basement rocks in the core of the anticline. Access is by 4-wheel drive from Rustaq, westwards and then south into the gorge. The gorge profile cuts down section from north to south from the Cretaceous Natih Formation down to the Permian Saiq Formation. Some small-scale intra-formational thickening occurred by internal duplexing locally along the north flank of Jebel Akhdar (Breton et al. 2004). These are very small-scale folds and thrusts (Fig. 10.23), restricted to the Cretaceous part of the shelf sequence and are not continuous up to the underlying shelf or overlying allochthonous units. These duplexes are not regional north-verging folds, but local features related to thickening along the flanks of the main Jebel Akhdar fold. The Middle Permian unconformity is spectacularly exposed all around the Window (Figs. 10.24 and 10.25). South of the unconformity the high and steep sided mountains of the shelf carbonates give way to low hills and open ground of the Sahtan bowl floored by Precambrian Muaydeen (+Shuram) Formation siltstones.

In the Sahtan bowl several smaller roads branch off to the west, firstly to the village of Al Hajir, then another to Madruj and a third to Al Hub (Figs. 10.26 and 10.27). There are superb steep and technical rock climbing—trekking routes up to the Jebel Shams summit ridge from each of these villages. A new road from the upper Wadi Sahtan from near Madruj to the village of Yasab has opened up the western part of the Jebel Shams ridge (Fig. 10.28). The main Wadi Sahtan road extends south to the village of Wajmah from where the main climbing route up to Jebel Shams starts (Figs. 10.29 and 10.30). This road has stunning views along the north face of Jebel Shams and across the Sahtan bowl. The villages are all in incredible settings nestling beneath giant cliffs and overhangs with the tiered limestone cliffs of the 2000 m high Jebel Shams north face looming above. The views from the summit of Jebel Shams are superb, especially at dawn (Fig. 10.31).

10.8 Jebel Shams North Face

The most spectacular part of the entire mountain belt is the North face of Jebel Shams reaching an altitude of 3009 m high at the main summit and 3005 m at the North summit. The Jebel Shams ridge runs west to east along the main watershed of the Oman Mountains and extends for over 50 km (Fig. 10.32). Four main trekking and climbing routes ascend the cliffs to the Jebel Shams summit starting at the villages of (a) Madruj, (b) Al Hub, (c) Wajmah and (d) Bani Habib. These four routes are long two-day climbs with more than 2000 meters of ascent and some technically difficult climbing (particularly on the Madruj and Al Hub routes). Route finding can be difficult with several long traverses along high ledges required to avoid steep cliff sections. All four routes offer outstanding rock

Fig. 10.22 Water-worn pavement of dark limestone with abundant coiled gastropods, Early Cretaceous Habshan Formation, Wadi Nakhr

climbing expeditions, ascending the entire Permian and Mesozoic shelf carbonate sequence with wonderful views in all directions.

The southern slopes of Jebel Akhdar are a gentle 20–30° dip-slope (Fig. 10.33), whereas the north faces are sheer cliffs cascading down in layers of cliff-forming limestones. The entire pre-Permian basement and Middle Permian to Cretaceous shelf carbonate sequence is magnificently exposed in 2000 m high cliff faces. The main Jebel Akhdar anticline axis runs through the middle of the Sahtan bowl with the Jebel Shams ridge to the south. The anticline is slightly asymmetric in the west, with a steeper SW flank suggesting a possible hanging-wall anticline above a blind thrust deep in the basement. The steep SW flank can only be seen near Murri, a small village at the western 'nose' of Jebel Akhdar.

Fig. 10.23 Small north-vergent thrust duplexes exposed in Sahtan Group shelf carbonates, Wadi Sahtan. These are minor accommodation structures along the northern flank of Jebel Akhdar forming during culmination, probably in the Latest Cretaceous

10.9   Wadi Bani Awf, Snake Canyon

The upper section of Wadi Bani Auf (or Bani Awf) is the 'Snake Canyon', where spectacular recumbent folds in the Proterozoic sediments are seen in glorious 3-D. Snake canyon (Wadi Bimmah) downstream from the village of Bilad Seit (950 m) is probably the best canyoning route anywhere in Arabia (Fig. 10.34). The gorge cuts through strongly folded massive black sulphurous dolomites and limestones of the Precambrian Hijam (=Khufai) Formation and winds down through an amazing convoluted canyon (Fig. 10.35). The descent requires abseiling (rappelling) from fixed bolts into wadi pools with several sections requiring short swims. The route was first descended in 1981 and since then, bolts and fixed ropes have been emplaced. It is impossible to get out of the gorge in a hurry, so if the weather is at all rainy the whole canyon should be avoided. Flash floods are common and

Fig. 10.24 Palm groves and villages beneath the limestone cliffs on the north face of Jebel Shams, upper Wadi Sahtan

Fig. 10.25 North face of Jebel Shams showing the full geological section through the Middle Permian to Cenomanian shelf carbonate sequence

Fig. 10.26   Al Hub village, north face Jebel Shams from upper Wadi Sahtan

several parties have been trapped in the gorge without warning, with extremely
serious consequences. In 1996 nine hikers were drowned when a flash flood swept
down the gorge. The lower gorge has a second section of incredible narrow
deep-cut canyon with long swimming pools and steep to overhanging cliffs
(Fig. 10.36). The 'little snake gorge' presents an easier, safer and shorter alternative. The Hijam Formation was tightly folded in a pre-Permian event, possibly as
old as Cambrian (Fig. 10.37).

Bilad Seit is one of the most scenic villages in all the Oman Mountains that
used to be completely cut off from the rest of the country, accessible only by

Fig. 10.27 Looking up the limestones cliffs of the Al Hub climbing route, Jebel Shams

two-day donkey ride along Wadi Bani Auf. A new dirt road was bulldozed into
Bilad Seit in 1981, and another connecting over the Jebel Akhdar ridge followed,
so that now a complete circuit from north to south can be driven in a four-wheel
drive vehicle. It is possible to trek up the big cliffs to the south from Bilad Seit to
access the summit watershed ridge in the eastern end of the Jebel Shams ridge.
This is a complex route that weaves up the steep cliffs to a prominent saddle at

Fig. 10.28 The road to Yasab, southwestern flank of the Sahtan bowl. The Mid-Permian unconformity lies along the road, with cliffs of Saiq Formation towering above. *Note* Land Rover for scale on the road

2050 m and gives wonderful views along the entire mountain range. It is possible to trek along the ridge to the east then descend down a very steep wadi over 900 m to the tiny settlement of Hat (1150 m).

The central part of Wadi Bani Auf cuts down to the Precambrian rocks in the core of the Jebel Akhdar anticline. Spectacular folds in the Precambrian Hijam dolomite are exposed along the upper road to Bilad Seit (Fig. 10.37). Mu'aydeen Formation siltstones form a very prominent 'pencil cleavage' due to narrow spaced intersecting cleavage planes formed during the Paleozoic orogenic event.

Fig. 10.29 Old photographs (ca. 1979) of Wajma village, starting point of the main climbing route to Jebel Shams

10.10   Wadi Bani Kharus

Wadi Bani Kharus is the large deeply incised wadi that cuts through the shelf carbonates south of Awabi. Like Wadi Sahtan and Wadi Bani Auf it cuts through the entire shelf carbonate succession from the Natih Formation in the north through the entire Lower Cretaceous, Jurassic, Triassic and Upper Permian succession. The Mid-Permian unconformity is obvious as the narrow canyon opens out into the lower hills of the Precambrian succession. Beneath the unconformity Late Precambrian limestones and dolomites with stromatolites of the Kharus Formation are approximately 650 million years old (Fig. 10.38). Above the unconformity, a thin band of conglomerates is overlain by the massive, cliff-forming dolomites of the Middle Permian Saiq Formation deposited 260–250 million years ago as Arabia rifted away and developed a passive continental margin. Thus, 400 million years of time is missing along this sharp unconformity.

Fig. 10.30 Wajma village in Wadi Sahtan, starting point of the climb up the North face of Jebel Shams

Some of the best exposures of the Neoproterozoic–Cambrian glacial deposits occur along Wadi Bani Kharus and a subsidiary Wadi Haslam (Wadi Hajir) as well as along Wadi Mistal to the east. The Snowball Earth theory invokes periods of global glaciation with frozen oceans followed by a runaway greenhouse world where 'cap carbonates' were deposited above glacial deposits. In Oman, at least two major glacial events are recorded in tillites of the Ghubrah and Fiq Formations

Fig. 10.31 View east along the Jebel Shams ridge at dawn from the summit of Jebel Shams, 3009 meters

Fig. 10.32 View west along the main Jebel Shams ridge above the village of Bilat Seit. Lower hills to the north are the Hajir dolomites in the Precambrian basement

Fig. 10.33 Old photograph (ca. 1977) of the southern slope of Jebel Akhdar with date palms around Al Hamra in the foreground

Fig. 10.34 Bilad Seit village, upper Wadi Bani Auf with the shelf carbonate cliffs of Jebel Akhdar beyond

Fig. 10.35 Snake gorge, Wadi Bani Auf, cutting through fetid black dolomites of the Precambrian Hajir Formation. Snake gorge is one of the best canyoning routes in Arabia

and the cap carbonate is present as the Hadash Formation (Fig. 10.39). An ash horizon in the lower part of the Ghubrah Formation (Abu Mahar Group, Huqf Supergroup) has a U-Pb zircon age of 711.8 million years making these rocks equivalent to the Sturtian glaciation of the Neoproterozoic (Allen 2007). The Fiq Formation glacial rocks are related to the younger Marinoan glaciation. Paleomagnetic data show that both the glacial tillites and the cap carbonates formed in tropical latitudes giving good evidence that glacial-interglacial cycles occurred at low latitudes. All these rocks can be seen in Wadi Bani Kharus, Wadi Bani Auf and Wadi Mistal (Leather et al. 2002; Kilner et al. 2005).

Three excellent treks ascend from Wadi Bani Kharus to the northern rim of the Saiq plateau (Dale and Hadwin 2001). One route starts in the village of Thaqib and climbs approximately 1300 m to a tiny ruined mosque at Talhat. Another starts at the delightful village of Al'Ulyah and ascends the so-called 'Persian Steps' to gain a large forested plateau area full of juniper trees and wild olive trees. At about 2000 m altitude, the tiny remote village of Saqrah, now deserted, nestles along a 20°

Fig. 10.36 Pools in the lower canyon Wadi Bani Auf

dip-slope tucked beneath overhangs in an incredible position. From here it is possible to link east and south towards Saiq and Shurayjah, or to contour west towards Wadi al Khawar and Wadi al-Hijri.

The SAS route climbs from the village of Al Hajir, west of Wadi Bani Kharus, past the small green oasis of Halhal at 1400 m up to the Saiq plateau. The route is named after the British army unit that stormed the Jebel Akhdar stronghold of the Imam Sulaiman bin Himyar during the jebel war in the 1950s. They drove the rebel

Fig. 10.37 Tight folds in the Precambrian Hajir Formation, Wadi Bani Auf in the core of the Jebel Akhdar anticline

tribesmen down from Jebel Akhdar and after their village of Tanuf was bombed the Imam's followers were driven into the desert and the whole Jebel Akhdar region became united with the Sultan.

10.11 Wadi Mistal and the Ghubrah Bowl

Half way between the towns of Nakhl and Awabi on the road to Rustaq, a turnoff to the south follows a deep cut gorge of Wadi Mistal (Fig. 10.40). This road traverses from the mantle sequence peridotites of the Semail ophiolite across a

Fig. 10.38 Middle and Upper Permian Saiq Formation limestones unconformably above Buah Formation glacial sediments, Wadi Bani Kharus; photo courtesy of Bruce Levell

Fig. 10.39 Hadash Formation cap carbonates on Fiq Formation Precambrian glacial sediments, Wadi Bani Kharus; photo courtesy of Bruce Levell

Fig. 10.40 Wakhan village, Wadi Mistal, view northwest towards Bani Auf and the Sahtan bowl

major normal fault into the shelf carbonates and beyond to the Precambrian rocks of the Ghubrah bowl. A right hand fork in the road leads to the village of Al Hijar, and further to the high village of Wakan. From Wakan several trekking routes lead up into the high mountains to the south and east, and to access the eastern margin of the Saiq plateau via a track to the remote village of Al Manakhir. It is possible to trek across the watershed and connect to Wadi Halfayn draining east towards the Semail Gap. There is also an excellent trek between Wakan village and Hadash.

Back in the Ghubrah bowl the main road continues east to dead-end at the wonderful green and pleasant village of Hadash, surrounded by spectacular high limestone mountains of Jebel Nakhl (Fig. 10.41). Hadash is renowned for its excellent rock climbing and bouldering on enormous fallen blocks of limestone. It is also the type locality for the Hadash Formation, the Marinoan glacial cap carbonate.

Fig. 10.41   Aerial photograph of the Jebel Akhdar anticline

10.12   Wadi Halfayn and Semail Gap

The road from Muscat to Ibri and Nizwa along the Semail Gap winds through the ophiolite hills south of Fanjah and enters a complex area of imbricated meta-morphic sole rocks (greenschist and amphibolite facies rocks forming during ophiolite obduction by heat and shearing along the base of the ophiolite). The main road then follows the Semail Gap, with the 2000 m cliffs of Jebel Nakhl to the west and the lower dark-coloured hills of the ophiolite mantle sequence to the east. The original contact was a south-directed thrust with ophiolite thrust over metamorphic sole over Hawasina–Haybi complex allochthous slices and all of these over the Cretaceous shelf carbonates of Jebel Nakhl–Akhdar. During the latest stages of thrusting the culmination of Jebel Akhdar–Nakhl anticlines was accompanied by late-stage thrusting in the pre-Permian basement. The Semail Gap is a lateral ramp associated with this late-stage culmination of Jebel Akhdar–Jebel Nakhl–Saih Hatat fold. Late-stage normal faulting may also have occurred along the Semail Gap because the ophiolite peridotites rest nearly directly on the Cretaceous shelf carbonates. Late stage deep thrusting and culmination of Jebel Akhdar–Jebel Nakhl probably occurred during the final stages of the late Cretaceous ophiolite emplacement event although further growth of the Jebel Akhdar culmination (possibly between 2 and 3 km of uplift) occurred after the Late Eocene–Early Miocene). Correlations with the Abat basin, south of Sur, suggest that it is possible that this folding and uplift continued into the Pliocene–Pleistocene times. Along the Semail Gap several new road-cuts expose highly contorted red radiolarian cherts of the Hawasina complex showing spectacular chevron type folds.

10.13   Hawasina Foreland Fold-Thrust Belt, Jebel Nihayda, Hamrat Ad Duru Range

Structurally beneath the Semail Ophiolite, exposed around Nizwa and Bahla, a series of thin-skinned thrust sheets of Hawasina complex sedimentary rocks are exposed in the Hamrat ad Duru range. The Hawasina complex rocks have been thrust over the autochthonous shelf carbonates from NE to SW during the Late Cretaceous ophiolite obduction event. The Hamrat Duru Group forms the Mesozoic proximal slope facies sediments that are time-equivalent units to the shelf carbonates. They form a typical imbricated fold-thrust belt with a series of structurally repeated thrust slices of shales, sandstones, limestones, and cherts. These rocks include the Triassic Zulla Formation (limestones, sandstones, radiolarian cherts; *Halobia* limestone at top) the Jurassic Guwayza Formation (sandstones, oolitic limestones), and the Late Jurassic–Early Cretaceous Sidr Formation (mudstones overlain by cherts) The youngest rocks are the Albian-Aptian Nayid Formation limestones. These rocks have been described in detail by Glennie et al. (1973, 1974), Cooper (1988, 1990), Béchennec et al. (1990), Blechschmidt et al. (2004). The Hamrat ad Duru ranges are a maze of small linear jebels showing imbricate repetitions of the stratigraphy. Detailed mapping has resulted in restoration of these imbricate slices that gives a minimum amount of crustal shortening in these proximal Tethyan basin sedimentary rocks (Cooper 1988).

10.14   Jebel Nakhl

Jebel Nakhl (2138 m) is the prominent mountain range running NNE-SSW that parallels the western margin of the Semail Gap. It is formed of tightly folded shelf carbonates that are part of the Jebel Akhdar range. The main west-east axial plane of the Jebel Akhdar anticline swings abruptly at right-angles into the NNE trending Jebel Nakhl trend (Fig. 10.41). At the far northern end of Jebel Nakhl the same fold axis swings back east and plunges east beneath the Fanjah saddle. In this region all structural levels can be seen from the Precambrian basement rocks in the Ghubrah bowl, the Permian and Mesozoic shelf carbonates and the overlying Hawasina, Haybi and Semail ophiolite sheets around Fanjah. The trek to the summit of Jebel Nakhl starts at Nakhl village with a short scramble over a low pass into Wadi Hedik. The route follows this wadi all the way up past deep wadi gorges, steep slopes and airy traverses, with an unusual grassy meadow at about 1000 m altitude (Dale and Hadwin 2001). The views from the summit of Jebel Nakhl north across the Batinah coastal plain south and west to the Saiq plateau and Jebel Akhdar, and east over the ophiolite hills around Semail, are spectacular. It is also possible to ascend Jebel Nakhl from near Fanjah village in the Semail Gap.

10.15   Semail Gap

The Semail Gap is the lineament of the Semail thrust, reactivated as a down-to-east normal fault during the uplift of Jebel Akhdar–Jebel Nakhl shelf carbonates (Fig. 10.42). The geometry of the main anticline along the northern end of Jebel Nakhl suggests that some sort of basal detachment must exist in the pre-Permian

Fig. 10.42 Eastern flank of Jebel Nakhl along the Semail Gap view west

basement at depth. This detachment is likely to be similar to the one proposed at depth beneath both Jebel Akhdar and Saih Hatat. Both deep detachments have more displacement in the north than the south, and may ramp up to higher levels along a dominant kink-band set, as seen, for example, in the SW flank of Jebel Akhdar and the Wadi Tayyin area of southern Saih Hatat. The Semail Gap fault would therefore be related to a deeper-level lateral ramp at right angles to the main transport direction. Up to 3–4 km of throw occurs along this fault, which downthrows ophiolite mantle sequence rocks against structurally deeper Wasia Formation shelf carbonates.

The Semail Gap fault cannot be a strike-slip fault because it does not laterally offset any geology, and does not extend south into the foreland. Instead it is an S-shaped transfer between the Jebel Akhdar and Saih Hatat domes. The fault is most likely a listric normal fault curving around the Wasia Formation top-shelf, decreasing in throw to the north and south away from the central sector. The lateral ramp would therefore have been active during the latest part of the Late Cretaceous tectonic event and enhanced by some unquantifiable amount of late Cenozoic uplift. Prominent NNE-trending fold axes also occur along the Ba'id corridor, where the Semail ophiolite has been folded around a N-S-aligned fold axis, along the NNE-aligned Jebel Qirmadil and the Wadi Mayh sheath folds around the northern flank of the Saih Hatat culmination (Chaps. 11 and 12).

References

Allen PA (2007) The Huqf Supergroup of Oman: Basin development and context for Neoproterozoic glaciation. Earth Sci Rev 84:139–185

Béchennec F, LeMetour J, Rabu D, Bourdillon-De-Grissac C, DeWever P, Beurrier M, Villey M (1990) In: Robertson AFH, Searle MP, Ries AC (eds) The geology and tectonics of the Oman region. Geological Society, London, Special Publication no. 49, pp 213–224

Blechschmidt I et al (2004) Stratigraphic architecture of the northern Oman continental margin–Mesozoic Hamrat Duru Group Hawasina complex, Oman. GeoArabia 9:81–132

Breton J-P, Bechennec F, LeMetour J, Moen-Maurel L, Razin P (2004) Eoalpine (Cretaceous) evolution of the Oman Tethyan continental margin: insights from a structural field study in Jabal Akhdar (Oman Mountains). GeoArabia 9(2):41–58

Cooper DJW (1988) Structure and sequence of thrusting in deep-water sediments during ophiolite emplacement in the south-central Oman Mountains. J Struct Geol 10:473–485

Cooper DJW (1990) Sedimentary evolution and palaeogeographical reconstruction of the Mesozoic continental rise in Oman: evidence from the Hamrat Duru Group. In: Robertson AFH, Searle MP, Ries AC (eds) The geology and tectonics of the Oman region. Geological Society, London, Special Publication no. 49, pp 161–187

Dale A, Hadwin J (2001) Adventure trekking in Oman, 253 p

Droste H (2005) Stratigraphic evolution and stratal geometries of the Cretaceous carbonate platform in Oman. Geological Society of Oman, Field Guide 014

Filbrandt JB, Nolan SC, Ries AC (1990) Late Cretaceous and early Tertiary evolution of Jebel Ja'alan and adjacent areas, NE Oman. In: Robertson AFH, Searle MP, Ries AC (eds) The geology and tectonics of the Oman region. Geological Society, London, Special Publication no. 49, pp 697–714

Glennie KW, Boeuf MG, Hughes-Clarke MHW, Moody-Stuart M, Pilaar WF, Reinhardt BM (1973) Late Cretaceous nappes in the Oman Mountains and their geologic evolution. Bull Am Assoc Pet Geol 57:5–27

Glennie KW, Boeuf MG, Hughes-Clarke MHW, Moody-Stuart M, Pilaar WF, Reinhardt BM (1974) Geology of the Oman Mountains. Verhandelingen Koninklijk Nederlands geologisch mijnbouwkundidg Genootschap no. 31, 423 p

Holland M, Urai JL, Muchez P, Willemse E (2009) Evolution of fractures in a highly dynamic, thermal, hydraulic, and mechanical system (1) Field observations in Mesozoic carbonates, Jabal Shams, Oman Mountains. GeoArabia 14:57–110

Hughes-Clarke M (1990) Oman's geological heritage. Petroleum Development Oman Ltd, 247 p

Kilner B, MacNiocall C, Brasier M (2005) Low-latitude glaciation in the Neoproterozoic of Oman. Geology 33:413–416

Leather J, Allen PA, Brasier MD, Cozzi A (2002) A Neoproterozoic snowball earth under scrutiny: evidence from the Fiq glaciation of Oman. Geology 30:891–894

Schreurs G, Immenhauser A (1999) West-northwest directed obduction of the Batain Group on the eastern Oman continental margin at the Cretaceous-Tertiary boundary. Tectonics 18:148–160

Searle MP (2007) Structural geometry, style and timing of defrmation in the Hawasina window, Al Jabal al-Akhdar and Saih Hatat culminations, Oman mountains. GeoArabia 12:93–124

Van Buchem FSP et al (2002) High-resolution sequence stratigraphic architecture of Barremian/Aptian carbonate systems in Northern Oman and the United Arab Emirates (Kharaib and Shu'aiba Formations). GeoArabia 7(3):461–498

Wilson HH (1969) Late Cretaceous eugeosynclinal sedimentation, gravity tectonics and ophiolite emplacement in the Oman Mountains. Am Assoc Pet Geol 53(3):626–671

Muscat–As Sifah

11

Muscat is the only capital city in the World to be built on peridotite mantle rocks. The capital area around Muscat includes a wide variety of spectacular geological sites and geomorphological settings (Fig. 11.1). The great mountain ranges of Jebel al-Akhdar and Jebel Nakhl come down to the coast as the Batinah coastal plain ends in the mangrove creeks of Qurum and the mouth of Wadi Aday. The Semail ophiolite blocks around Rustaq extend beneath the gravel plains of Azaiba to the north (Fig. 11.2). At Fanjah great wadis flowing north from the Semail Gap carve a sinuous passage between the high limestone massifs of Jebel Nakhl to the west and Saih Hatat to the east. Late Campanian–Maastrichtian conglomerates infill fault-bounded basins related to the rising Oman Mountains to the south and a deepening hinterland basin developing offshore. Highly fossiliferous Paleocene-Eocene limestones, deposited unconformably above the ophiolite and all underlying units, have been folded around huge anticlines with north-south axes at right angles to the regional trend of the Mesozoic shelf carbonates inland. These limestones are beautifully exposed around Qurum and Ras al Hamra, and are folded over two large-scale anticlines with axes running NNW-SSW at Mina al-Fahal and across the Muscat–Muttrah peridotite. These Paleogene limestones outcrop around the Bandar Jissah and Bandar Khuyran areas to the east where they form a spectacular drowned coastline (Fig. 11.3).

The drowned coastline from the Muscat area to As Sifah beach is rugged and beautiful, with mountains descending right to the sea. Much of the coastline is accessible only by sea or long treks over the hills from the south. New roads are being constructed all along the northern coastal region with a major road winding through the hills inland from Bandar Khuyran to As Sifah. Numerous small white sand beaches are scattered along this stretch of coast with narrow fjords testifying

© Springer Nature Switzerland AG 2019
M. Searle, *Geology of the Oman Mountains, Eastern Arabia*,
GeoGuide, https://doi.org/10.1007/978-3-030-18453-7_11

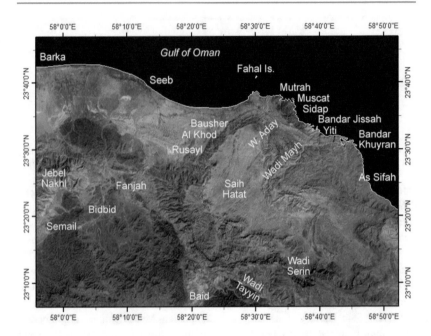

Fig. 11.1 Landsat photograph of the Muscat–As Sifah area showing key locations

to a recently drowned origin, due to northward tilting of the continental margin. Geologically this region is of the utmost importance as it shows the only location in the World where high-pressure eclogites related to deep subduction (at As Sifah) are preserved at deep structural levels beneath an obducted ophiolite complex (Muscat–Muttrah ophiolite). A completely exposed profile across a Cretaceous subduction zone more than 20 km thick is exposed from As Sifah up to Muscat showing eclogites, blueschists and carpholite-grade high-pressure metamorphic rocks (Fig. 11.4). All these rocks were intensely deformed during prograde burial to ~ 23 kbar (>100 km depth) in a subduction zone setting and were subsequently exhumed back along the same subduction channel. Isoclinal folds can be seen throughout the section and the World's largest sheath fold (Wadi Mayh sheath fold) is spectacularly exposed along the northern margin of Saih Hatat (see Chap. 11).

Fig. 11.2 Landsat photograph of the Saih Hatat region and Muscat to As Sifah region

Fig. 11.3 Paleocene–Eocene limestones around Bandar Jissah lie unconformably above the peridotites around Muscat in distance

Fig. 11.4 Digital elevation model of the Muscat–As Sifah–Saih Hatat region showing major tectonic features

The entire coastal region around Muscat and Muttrah, between Qurum–Ras Al Hamra and As Sifah, south to Wadi Mayh has many very important geological localities, showing unique and spectacular geology, geomorphology and natural history. Inevitably this region is becoming highly developed but future planning must allow for these sites of special scientific interest and beauty to be preserved.

11.1 Muscat Ophiolite

The capital city of Oman, Muscat, must be one of the most spectacular settings of any capital in the World (Figs. 11.5 and 11.6). The small city is completely surrounded by jagged black hills of mantle sequence rocks, harzburgite and dunite, with every prominent peak having a small mud-brick watchtower. Two impressive large forts, Mirani and Jalali guard the entrance to Muscat harbour. Along the east side of Muscat harbour an island, Jazirat Muscat, beyond Fort Jalali is cut off

Fig. 11.5 Muscat, capital city of Oman, surrounded by peridotite hills with Fort Mirani and the Sultan's palace

Fig. 11.6 Muscat island beyond the Fort Jalali build on harzburgites of the ophiolite mantle sequence

during high tides. Important habitats for corals, shells and fish are present all along the eastern shore of Muscat Island. Southeast of Muscat the sea-cliffs continue to a tiny bay, Cemetery Bay, containing the graves of several unfortunate western sailors and missionaries. Further south is the busy fishing village of Sidap, previously sited on a beautiful white sand beach, but now a busy harbour. Two further beaches south of Sidap are Harambol, also a tiny fishing village, and Bandar Rawdah now the site of a yacht club and harbour. A small island and sea stack at Cat Island 100 m offshore is ringed by beautiful coral reefs. This entire coastline hosts unique hard and soft corals which are home to numerous species of cowrie, including the famous *Cypraea pulchra* the 'four eyes' cowrie that lives on branching corals along this coastline. Schools of fish are seen on every dive and occasional eagle rays and reef sharks quarter the reef. Offshore, it is not uncommon to see schools of hundreds of dolphins, manta rays leaping out of the water, and the occasional breaching whale. Hills of peridotite drop straight into the sea with beautiful sandy beaches in-between, extended south to Al Bustan beach (Fig. 11.7). Paleocene–Eocene limestones unconformably overlie the harzburgites of the Muscat ophiolite and at Al Bustan they are tilted steeply to the south adjacent to the Wadi Kabir fault (Fig. 11.8).

Fig. 11.7 a Aerial photograph of Muscat harbour with Fort Jalali, the Sultan's palace and Fort Mirani. **b** Old photograph (ca. 1976) of Al Bustan beach south of Muscat

Fig. 11.8 Paleocene–Eocene limestones, steeply tilted to the south, lying unconformably on harzburgites of the mantle sequence of the Muscat ophiolite, Al Bustan

11.2   Muttrah Harbour Banded Peridotites

The corniche road that runs along Muttrah harbour has a spectacular backdrop of spiky black ophiolite hills surrounding the port area to the west and the Muttrah fort area to the east (Fig. 11.9). Muttrah town spreads inland towards the Ruwi valley and a few of the harzburgite mountains have been bulldozed to make room for new buildings. The hills beyond Muttrah fort show impressive outcrops of the 'banded ultramafic unit' along the base of the ophiolite. These hills are composed of alternating layers of pale brown dunites with darker brown harzburgites. The dunite layers have been interpreted as relic melt pathways through which gabbroic

Fig. 11.9   Muttrah harbour and fort built on harzburgites and dunites of the ophiolite mantle sequence

melt has been extracted leaving a restitic dunite. These ultramafic rocks are riddled with characteristic 'fox holes' eroded out of serpentinite fractures. The banded ultramafic unit shows evidence of mylonitic shearing fabrics formed during emplacement of the ophiolite onto the Arabian continental margin. The ultramafic rocks have been strongly serpentinised and thin veins of green serpentine, asbestos, talc and white magnesite are common. It is possible to scramble up some of the wadis behind the fort and along the corniche road, over extremely loose rocks. The hilltops provide excellent views over the Muscat-Muttrah peridotites and the ocean beyond. Inland Maastrichtian Qahlah Formation and Paleocene limestone unconformably overlie the Muttrah peridotites (Fig. 11.10).

Fig. 11.10 Muttrah corniche surrounded by hills of peridotite with unconformable Paleocene limestones above

11.3 Darsayt Ophiolite Section

A complete section through the ophiolite can be accessed from the village of Darsayt westwards along the shore (Fig. 11.11). The unconformity with Paleocene limestones overlying Mantle sequence peridotites is beautifully exposed along road cuts and the mountains between Saih al Maleh, or Mina al Fahal (the PDO office beach), and Darsayt village to the east. The harzburgites of the Mantle sequence in turn overlie the Ruwi mélange which occupies most of the Ruwi valley. The Ruwi mélange consists of a complex mixture of blocks including cherts, lawsonite-bearing basalts, dolomites and limestones in a shaly matrix that contains the high-pressure mineral carpholite. From Darsayt a full ophiolite sequence can be traced towards the west with gabbros passing up to sparsely exposed sheeted dykes

Fig. 11.11 Darsayt beach and Muttrah harbour in the distance surrounded by peridotite hills of the Semail ophiolite

and pillow lavas. Pillow lavas also outcrop on the western limb of the Mina al Fahal anticline at PDO Gate 3, and sheeted dykes are exposed next to Gate 2. The ophiolite sequence has been tilted and eroded after emplacement such that the overlying Palaeocene limestones rest unconformably on all units of the ophiolite.

The unconformity beneath the Maastrichtian and Cenozoic rocks is undulating and it is clear that there was considerable topography following ophiolite emplacement. Immediately above the unconformity thin lateritic paleo-soil horizons in a few places, such as Ghala, are overlain and interbedded with conglomerates of the Campanian–Maastrichtian Al Khod Formation infilling large erosional gullies. Clasts within the Al Khod Formation include ophiolite and chert boulders as well as limestones and dolomites of the shelf carbonates and pre-Permian quartzites. This full set of erosional debris shows that the Oman Mountains were sub-aerial at this time, folded and topographically high, the source for wadi systems draining northwards towards more open deeper water deposits (Thaqab

Formation exposed further north along the Batinah coast). In the Darsayt–Mina al Fahal area, the Late Paleocene–Early Eocene Jafnayn Formation limestones lie unconformably on the basal conglomerates and the ophiolite beneath. These rocks are yellow marly limestones rich in foraminifera (*Alveolina*, *Milliolida* sp.), echinoids, bivalves, coral fragments and gastropods (e.g. *Turitella* sp.). Impressive large veins of gypsum cut through the limestones along the coastal section around the headland to Mina al Fahal.

11.4   Ruwi Valley

The Ruwi valley is underlain by a complex tectonic mélange called the Ruwi mélange which underlies the ophiolite rocks exposed around Muscat and Muttrah and overlies shelf carbonates (Fig. 11.12). It is comprised of blocks of cherts, limestones, dolomite and meta-basalts that contain the high-pressure mineral lawsonite in a shaley matrix that contains the high-pressure mineral carpholite. A single isolated huge block of Sahtan Group shelf carbonates showing a

Fig. 11.12 Aerial photograph of the Ruwi valley, view looking southeast towards Bandar Jissah, Bandar Khuyran and the mountains around Wadi Mayh

Fig. 11.13 Ruwi valley floored mainly by high-pressure Ruwi mélange with various sedimenatry lithogies and lawsonite basalt blocks in a carpholite bearing mudstone matrix. One large block of isoclinally folded Sahtan Formation Cretaceous shelf carbonate crops out on the northern side of the valley. In distance are white hills of Paleocene limestones unconformably overlying Ruwi mélange and ophiolitic rocks around Darsayt

spectacular large-scale recumbent fold can be seen in the mountainside to the north of the southern part of the Ruwi valley (Fig. 11.13). The Ruwi mélange is interpreted as a trench mélange, equivalent to the Haybi complex mélanges immediately beneath the Semail ophiolite. The Ruwi mélange extends westwards to the Bausher area where blocks of serpentinised peridotites occur and eastwards as far as Bandar Jissah, where it underlies the unconformable Paleocene-Eocene limestones.

The road from Ruwi east to Qantab and Bandar Jissah follows a major fault along Wadi Kabir where it can be demonstrated that the latest normal fault (top-north) motion was post-Paleogene. The hills to the south are comprised of Triassic Mahil Formation shelf carbonate overlain unconformably by rubbly Eocene limestone dipping at 20° S. These limestones have been dropped down to the north by over 100 m. The Paleocene limestones to the north however lie unconformably on top of peridotites of the Muscat ophiolite. The Wadi Kabir fault almost certainly had a Late Cretaceous origin during exhumation of the high-pressure carpholite-bearing shelf carbonates to the south, but was reactivated after the Lower Eocene during development of the Muscat–Muttrah fold to the north. Both normal faults were probably formed during periods of crustal compression, reflecting differential uplift rather than crustal extension.

11.5   Qurum–Ras Al Hamra

The coastal Batinah plain ends in the south at Qurum where the mountains abruptly come down to the sea. Paleocene and Lower Eocene shallow marine limestones exposed along the north coast between Qurum and Bandar Khuyran represent the post-ophiolite obduction sediments that were deposited unconformably above the Mesozoic shelf carbonates and all the overlying thrust sheets. A basal conglomerate unit, the Qahlah Formation is Late Campanian–Maastrichtian in age and variable in both composition and thickness, infilling fault-bounded grabens. Normal faults were active during this period controlling deposition in basins bounding the Oman Mountains to the south. Several of these faults have been traced offshore on seismic lines controlling deep Late Cretaceous basins infilled with a deeper water facies (Thaqab Formation).

Excellent sections through the mainly Paleocene shallow-marine limestones are seen around Ras al-Hamra which unconformably overlie the ophiolite exposed in the core of the Mina al Fahal anticline (Fig. 11.14). The limestones are packed with larger foraminifera notably *Lockhartia*, *Kathina*, *Daviesina* and calcareous algae. Coral limestones, containing abundant *Miliolid* and *Nummulites* formanifera, echinoids and bivalves, and algal bioherms (*Ethelia*) make well-bedded white limestones exposed at Ras al Hamra. Middle Eocene limestones of the Seeb Formation are brown-weathering, bioturbated nodular limestones approximately 350 m thick and contain abundant foraminifera including *Nummulites*, orbitoids alveolinids, miliolids and algae (Nolan et al. 1990). The youngest thick limestones exposed along the northeast flank of the Oman Mountains are Eocene, possibly Oligocene. After this time during the Miocene–Pliocene the Oman Mountains experienced uplift and compressional deformation in the form of large open box folds seen along the whole of the northern coastal areas of the Oman Mountains as well as along the southwestern flank of the mountains from UAE south to Ibri. A steep normal fault is exposed on Ras al Hamra headland (Fig. 11.15).

Fig. 11.14 Western flank of the Ras al Hamra box fold, Paleocene Seeb Formation limestones

Fig. 11.15 Steeply dipping Upper Paleocene Seeb Formation limestones with a faulted contact at Ras al Hamra headland, Fahal Island in the distance

Also exposed around Ras al Hamra beach are aeolianites, or fossilised wind-blown sands of Quaternary age preserved in small dry valleys. One of these aeolianites from the Ras al Hamra headland, east of the main beach, was dated by OSL at 130 ka BP (Glennie 2005). The aeolianite has bioclastic carbonate particles cemented by a calcite matrix as well as a non-carbonate fraction composed of quartz that was probably derived from Saih Hatat. Glennie (2005) interpreted the aeolianite as having been cemented during the interglacial interstadial period of higher rainfall and humidity during the early Holocene.

11.6   Fahal Island

Fahal (Arabic: turtle) Island lying four kilometers north of Ras al Hamra is a prominent offshore landmark with steep cliffs surrounding the island on all sides and is only accessible by boat (Fig. 11.16). The island is made up of Paleogene limestones, and is riddled with narrow caves and passages at sea level. It is an important geological site but is even more important as an extremely valuable conservation area for wildlife, especially nesting sea-birds and underwater life. The

Fig. 11.16 Fahal Island offshore Ras al Hamra, composed of tilted Paleocene shallow marine limestones, with some Greater Flamingoes on Qurum beach in foreground

rare Sooty falcon (*Falco concolor*) breeds on the sea cliffs of Fahal Island during the summer months, as does the ubiquitous Osprey (*Pandion haliaetus*). Oceanic birds such as terns, petrels, shearwaters, boobies and the magnificent Red-billed Tropic bird (*Phaethon aethereus*) are frequently seen around the island. The seas around Fahal Island have the greatest variety of corals in the whole country including four coral species found nowhere else in Oman and seven not found elsewhere in the Muscat area (Salm and Baldwin 1991). Numerous dive sites, all around Fahal Island, are famous sites for Whale Sharks (*Chelonioidea sp.*). Green, Hawksbill, Loggerhead, Leatherback turtles, and several species of ocean-migrating whales (Humpback, Bryde's, Sei whales in particular) have all been seen off Fahal Island. Large schools of Spinner dolphins (*Stenella longirostsis*), Common dolphins (*Delphinus carpensis*) and Bottlenose dolphins (*Tursiops truncatus*) are regularly seen in the seas around Fahal Island, sometimes in schools of hundreds. The rare Oman butterfly fish (*Chaetodon dialeucos*) was described for the first time in 1989 from these waters (Salm and Baldwin 1991).

11.7 Daymaniyat Islands

The Daymaniyat Islands are a string of nine main uninhabited islands located approximately 18 km offshore Ras Sawadi on the Batinah coast and 39 km northwest of Ras al Hamra. The Daymaniat islands are composed of Miocene coral limestones uplifted by Pliocene folding that may have also affected many rocks in the Muscat—Qurum area, and offshore. The islands have magnificent white sandy coral beaches and small cliffs along the north coasts that drop off to 20–25 m depths offshore. The islands were made a marine conservation nature reserve in 1996, and visitors are restricted, especially during the turtle nesting season. About 250–300 rare, highly endangered Hawksbill turtles (*Eretmochelys imbricarta*) nest on Jazirat Kharabah, the eastern Daymaniyat Island, from March to June, as do the more common Green Turtles (*Chelonia mydas*).

The beautiful fish-eating Ospreys are common along the Oman coasts and they also breed on the offshore islands. The Daymaniyat Islands form an important nesting sites for sea-birds notably terns, boobies and rare ocean-going tropic birds. The western island, Jazirat Jun and some of the central Daymaniyats have important dive sites with numerous species of corals and rocky ledges encrusted with multi-coloured anemonies, sponges and soft corals. The seas around the Daymaniyats are known for schools of bottlenose dolphins (*Tursiops truncatus*), common dolphins (*Delphinus delphis*) and spinner dolphins (*Stenella longirostris*), sting-rays (*Plesiobatidae* sp.), eagle rays (*Myliobatidae* sp.), leopard sharks (*Triakis semifasciata*), zebra sharks (*Stegostoman fasciatum*), occasional plankton feeding whale sharks (*Rhincodon typus*). The incredible humpback whale (*Megaptera novaeangliae*) has also been seen uncommonly, but regularly, off the Daymaniat islands (Salm and Baldwin 1991).

11.8 Bandar Jissah

The drowned coastline between Qantab village, Bandar Jissah and Yiti is a scenically spectacular and wonderful coastline, unique in Arabia and the World (Figs. 11.17, 11.18 and 11.19). The Paleocene–Eocene limestones that make up the mountains and coastline here are full of fossil echinoids, sponges, bivalves, gastropods, corals and foraminifera. Several strata show almost intact fossilised reefs with large coiled gastropods and beautifully preserved echinoids. At Bandar Jissah an impressive Eocene olistostrome crops out in the mountains inland from

Fig. 11.17 Paleocene to Middle Eocene limestones of the Seeb Formation around Bandar Jissah, harzburgites of the Muscat peridotite in distance

Fig. 11.18 Folded Paleocene–Lower Eocene limestones, Bandar Jissah, old photo prior to building the Dive Center

Fig. 11.19 Paleocene limestones around Bandar Jissah; Bar al-Jissah headland beyond

the Dive Center and road to Bar al-Jissah. Here, a channel over 300 m deep was eroded into the Eocene limestones and infilled with sedimentary mélange containing blocks of local Eocene limestones, and more exotic blocks of cherts, marls, carpholite-bearing limestones and green sandstones. These clasts were derived from a variety of structural units including the allochthonous Hawasina complex, and the high-pressure limestones of the Yiti unit of the shelf carbonate sequence. The Eocene olistostrome crops out between the Wadi Qanu fault and shear zone to the south and the Wadi Kabir normal fault to the north. The structures around this area suggest a major phase of deformation during the Late Cretaceous exhumation of the subduction related high-pressure eclogites and blueschists structurally below, and late subsequent phases of folding and faulting during the post-Maastrichtian, early Paleocene, and post-Eocene periods. Cenozoic folding on a large scale is apparent from the anticlines at Mina al-Fahal and Muscat-Bandar Jissah (Fig. 11.20).

Numerous small coves with sandy beaches are indented into the coastline and many unique snorkelling and dive sites are present along the coast and offshore islets (Salm and Baldwin 1991). Between Qantab and Bandar Jissah are three or four magnificent isolated small white sand beaches, surrounded by white limestone cliffs (Fig. 11.21). Up until the 1990s herds of mountain gazelle and even the occasional Arabian Tahr were commonly seen on the hills above Bandar Jissah, but these have been driven away by recent developments. Two of the larger offshore islands have excellent snorkelling sites with a variety of coral reefs and schools of reef fishes. Reef sharks, rays and occasional turtles are seen in the shallow bays. Large-scale new developments in Bandar Jissah and three new hotels on the headland beach to the east (Bar al-Jissah) have now made public access far more difficult. The headland north of Bar al-Jissah has a prominent natural arch eroded under the limestone cliffs through which fishing boats can pass at high tide (Fig. 11.22). The headland itself has numerous caves and passages at or below sea-level and some well-preserved coral reefs. The geomorphology shows that this whole stretch of coast between Muscat and Bandat Khuyran is a drowned coastline caused by recent tilting to the NE.

11.9 Bandar Khuyran

The Paleogene limestone hills of Bandar Jissah continue east to the next main set of indented fjords along the northward tilted drowned coastline of northern Oman, Bandar Khuyran (Fig. 11.23). The limestones here contain abundant fossils and show beautifully preserved fossil reefs of Paleocene and Eocene age, with

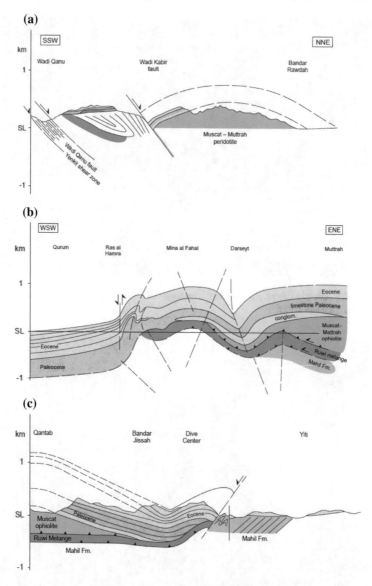

Fig. 11.20 Structural cross-sections across the Cenozoic folds of **a** the Muscat–Muttrah peridotites, **b** Mina al-Fahal, and **c** Bandar Jisaah structures

Fig. 11.21 Island in Bandar Jissah bay composed of Paleocene limestones. Modern coral reefs surround the island

Fig. 11.22 Natural arch eroded through Paleocene limestones, Bar al Jissah headline

Fig. 11.23 View southeast along the drowned coastline from Bandar Jissah towards Bandar Khuyran

branching corals, gastropods shells, bivalves and echinoids. The rocks are folded and tilted to the north and east and cut through by large-scale shear zones that separate the Paleogene limestone cover from the high-pressure rocks to the south (carpholite-blueschist and eclogite-facies). One of these shear zones is well-exposed inland at the western end of Bandar Khuyran and is mapped trending east across the bay, separating the outer islands from the inner creeks and inlets along the coast. The best section across the high-pressure zone from the upper levels around Bandar Khuyran structurally down-section to the carpholite- and blueschist-facies rocks down to the eclogites at As Sifah is the coastal traverse from Sand Dunes bay, east of Bandar Khuyran south to As Sifah, best seen from a boat.

The beautiful bays and inlets of Bandar Khuyran have small, rare sites of mangrove, and host coral reefs unrivalled in the Muscat area (Salm and Baldwin 1991). The creeks of Bandar Khuyran are very shallow and muddy and not suitable for beach-front hotel development. Two beautiful beaches northeast of Bandar Khuyran are Bommie Bay and Sand Dunes Bay, both with excellent snorkeling and diving. Here, coral reefs are well developed with many caves, inlets and rocky overhangs covered in luxuriant growths of soft corals and sponges. Turtles used to nest on these beaches but with an increasing amount of human interference many have already been driven away. The outer islands have deeper water and some beautiful snorkelling and diving sites. However, many of the inlets and bays of Bandar Khuryan are being seriously destroyed by visiting tourist boats, particularly with anchor damage to reefs, and litter.

11.10   Bausher, White Mountain

The northern margin of the Saih Hatat dome is a steep and prominent shear zone and normal fault that was active during culmination of the shelf carbonates. In places this shear zone and normal fault has reactivated or cut the earlier Semail thrust. At Bausher the contact is spectacularly preserved on White Mountain where a new road has recently been constructed linking the coastal plain across to the Saih Hatat bowl. White mountain is composed of a marble unit showing calc-mylonites along the contact zone. The presence of *Lithiotis* bivalve fossils indicate that the Bausher marble was part of the Lower Jurassic Sahtan Formation. Metamorphism probably occurred during ophiolite obduction in the Late Creta-ceous. The basal Semail thrust of the ophiolite in this area cuts up-stratigraphic section across the Jurassic to the Cretaceous part of the shelf carbonates from Bausher towards the south. Several north-vergent folds can be seen in the Jurassic shelf carbonates beneath the Bausher marble formed during uplift and culmination of Saih Hatat. The new road to Al-Amarat zig-zags up White Mountain and across the watershed to descend to Saih Hatat crossing progressively lower stratigraphic units of the Triassic Mahil Formation and Permian Saiq Formation towards the south.

North of the Bausher fault a narrow band of ophiolitic mélange is exposed around the base of White Mountain along the range east toward Ruwi, although much of this once spectacular geological outcrop has been quarried out and built upon. It consisted of large blocks of exotic white limestone-marble, red cherts and all components of the ophiolite enclosed in a serpentinite matrix, a classic example of an ophiolitic mélange. The overlying ophiolite is exposed only as a thin dis-continuous sheet above this, with extensive Paleogene limestones of the Jaffnayn, Rusayl and Seeb Formations exposed along the hills extending west to Rusayl and Al Khod. Excellent new exposures through these highly fossiliferous limestones can now been seen along the Muscat Expressway road-cuts.

11.11 Rusayl

The Rusayl embayment to the west of Bausher shows important structural constraints across the northern flank of Saih Hatat. Bedding and cleavage in the shelf carbonates are truncated abruptly at the flanking normal fault contact where the allochthonous units of the ophiolite, Haybi and Hawasina complex rocks have been down-faulted to the north. The normal fault probably relates more to the footwall uplift of Saih Hatat rather than any downfaulting of the ophiolite. Upper Palaeocene (Jafnayn 1 Formation carbonates) and Lower-Middle Eocene (Jafnayn 2, Rusayl, Seeb Formations) carbonates overlie peridotites, gabbros, sheeted dykes and pillow lavas of the Semail ophiolite. The ophiolite was already emplaced onto the Oman shelf, tilted to the NE, and partially eroded prior to deposition of the Late Maastrichtian (Al Khod and Simsima Formations) and Paleocene-Eocene limestones. The Paleogene limestones have been affected by broad open folding, dominantly of box-folding style, which must have begun after the Eocene, probably during the Oligocene-Miocene. The fold axes in the Paleogene rocks are aligned NNW-SSE and are not continuous across to the Mesozoic shelf carbonates. These structural relationships suggest that the Saih Hatat culmination had a long history from the original Cenomanian–Turonian thrusting of the ophiolite, cutting up-section to the south across the shelf carbonates, to uplift during the later part of the Late Cretaceous ophiolite obduction event. Late-stage deep thrusting in the basement probably caused uplift of Saih Hatat when the margin of the shelf carbonates was reactivated as a normal fault. A second phase of reactivation of the flanking normal fault occurred around a continuing uplifting Saih Hatat culmination.

11.12 Al Khod

The village of Al Khod (Al Khawd) lies north of Wadi Fanjah and shows a complete profile from the top of the Semail ophiolite. Approximately 800 m of Late Campanian–Maastrichtian conglomerates, interbedded with thin sandstones and shales, and rare limestones containing the Cretaceous microfossil *Loftusia* (Al Khawd Formation; Nolan et al. 1990) are lateral equivalents of the Qahlah Formation (Fig. 11.24). Surprisingly, many of the clasts in the conglomerate are quartzites, derived from the Ordovician Amdeh Formation in Saih Hatat as well as cherts and ophiolitic rocks. The lack of carbonate clasts from the shelf carbonates suggests that considerable structural complexity, topography and erosion occurred

Fig. 11.24 Road-cut through the Rusayl Formation and Al Khod conglomerates, near Al Khod village

prior to the Late Campanian. This phase of deformation is seen in the extreme folding, thrusting and shearing during exhumation of the High-pressure eclogites and blueschists of the North Oman High-pressure belt.

Following Late Cretaceous ophiolite obduction onto the previously passive Arabian continental margin, parts of the Oman Mountains just breached sea-level and terrestrial laterite deposits (tropical red-weathering soils) and deltaic sediments overlay the eroded ophiolite terrain. Some conglomerate bands contain vertebrate fossils, including rare dinosaur bones. A shallow marine transgression then drowned the area with highly fossiliferous white limestones containing abundant foraminifera as well as echinoids, gastropods and bivalves.

Sheared and highly serpentinised peridotites are exposed in a small quarry at Al Khod village, overlain by a thin band of mélange, correlated with the Batinah mélange. The overlying Al Khod Formation truncates all ophiolite units below and is overlain by another unconformity beneath the Paleocene Jafnayn, and Eocene Rusayl and Seeb Formations. The shallow-marine Paleocene limestones contain abundant foraminifera including *Nummulites*, sp., *Ranikothalia* sp., *Lockhartia* sp. and *Daviesina* sp., as well as abundant algal reefs. The Lower Eocene Rusayl Formation is marked by conglomerates, lagoonal mudstones, thin foraminifera and echinoid-bearing limestones and characteristic multi-coloured red, purple, orange and yellow shales. A spectacular road-cut through the multi-coloured rocks of the Rusayl Formation, cut by numerous steep faults can be seen approaching Al Khod village. Overlying the Rusayl Formation are shallow marine limestones of the Middle Eocene Seeb Formation containing abundant *Nummulites* sp. and *Assilina* sp. foraminifera.

These rocks have been folded along prominent box folds with steep flanks and minor thrusts developed along kink-bands. Near Sultan Qaboos University some outcrops of Middle Eocene Seeb Formation contain abundant foraminifera, notably *Nummulites* sp. and *Assilina* sp. These are single celled organisms forming flattened calcareous discs, and can reach an extremely large size, up to 5 or 6 cm across in this area. The foraminifera have been weathered out of the rock and lie on the surface in the thousands.

11.13 Fanjah

The main highway from Muscat to Nizwa leaves the Batinah coastal plain and winds through the ophiolite mountains towards Fanjah and Bid-Bid (Fig. 11.25). New road cuts near Fanjah show spectacular outcrops of the serpentinised harzburgites beneath the Maastrichtian–Paleocene unconformity. Some of these serpentinised peridotites have been completely altered by both silica-rich and calcium-magnesium-rich hydrothermal fluids. Silica serpentinites, called listvenites, are occasionally associated with gold deposits. Magnesium-rich hydrothermal fluids result in network of white veins of magnesite. In a few places brown-weathering laterites (palaeo-soils) indicate that parts of Oman were sub-surface during the Early Maastrichtian, prior to shallow water carbonate deposition over the eroded surface during the Late Maastrichtian–Paleocene. The brown-weathering basal serpentinites or listvenite mark the base of the Semail ophiolite (Fig. 11.26). Higher levels of the peridotite mantle sequence, the Moho transition zone and lower crust gabbros can be seen along Wadi Fanjah to the north, and along a new road cutting across the mountains to the town of Nakhl. South of Fanjah, imbricated thrust sheets of greenschists are part of the metamorphic sole. Thrust slices of garnet + diopside amphibolites are present in two main localities along the immediate base of the ophiolite east of Bidbid.

Fig. 11.25 Aerial photo view west over Fanjah village towards Jebel Nakhl

Fig. 11.26 Listvenite composed of silicified serpentinite, a result of metasomatic fluids fluxing along the Semail thrust, at Fanjah. Maastrichtian sediments with fossil dinosaur bones overlie the altered ophiolite

11.14 As Sifah Eclogites

The As Sifah eclogite locality is the only place in the World where eclogite facies metamorphic rocks occur in continental margin sedimentary rocks (mainly Permian Saiq Formation limestones and dolomites) beneath an oceanic upper plate (Semail Ophiolite and present day Gulf of Oman crust). The eclogite is the deepest structural level exposed in the Oman Mountains and part of a regional High-Pressure terrane that includes overlying retrogressed eclogites, blueschists and carpholite-bearing meta-sediments that extends south into the Saih Hatat basement (Fig. 11.27). The As Sifah eclogite locality has been described in detail by Searle et al. (1994, 2003, 2004) and Miller et al. (2002).

The As Sifah eclogites are exposed on a beautiful beach north of As Sifah village (Fig. 11.28). The eclogites include meta-basalts with the assemblage: garnet + clinopyroxene + glaucophane + phengite + epidote (Figs. 11.29 and 11.30) and

Digital Elevation Model of the Saih Hatat culmination, SE Oman Mountains

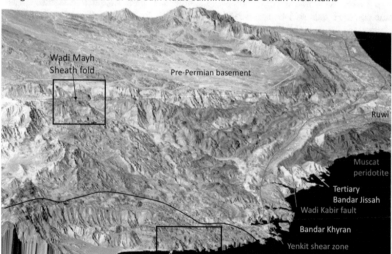

As Sifah eclogite GeoPark

Fig. 11.27 Digital elevation model of the Muscat–Sifah and northern Saih Hatat region

Fig. 11.28 As Sifah eclogite beach, showing the dark coloured eclogites beneath pale-coloured calc-schists and marbles, thought to be metamorphosed Permian Saiq Formation

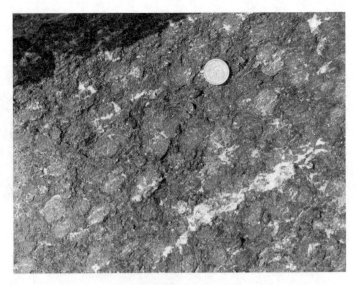

Fig. 11.29 As Sifah eclogite composed of dark red garnet, green clinopyroxene, dark blue-black glaucophane amphibole and white phengite mica

Fig. 11.30 Blue bladed glaucophane amphiboles in As Sifah eclogites

meta-pelites containing the assemblage: garnet + chloritoid + phengite + clinopyroxene. Meta-basic eclogites are derived from basaltic sills that intruded into Permian Saiq Formation limestones along the Oman continental margin and were subducted to depths of over 80–100 km during the latest stages of ophiolite obduction. The eclogites reached peak HP metamorphic conditions of 540 °C and 21–23 kbar at 79 Ma (Searle et al. 1994, 2004; Warren et al. 2005). P-T conditions are just below the quartz–coesite phase transition, but radial cracks in garnet indicate the presence of coesite at the greatest depths of burial. Excellent microstructural fabrics give clear indication of SW-directed exhumation of the high-pressure rocks with apparent NNE-directed extensional shear fabrics along the hanging-wall (Fig. 11.31). The eclogites occur as large- and small-scale scale boudins within calc-phengite schists (metamorphosed Permian Saiq Formation) and quartz schists (metamorphosed Amdeh, or possibly Al Khlata Formation) and in places show tight to isoclinal folds and sheath folds (Fig. 11.32). Above the eclogite boudins, highly sheared and deformed calc-schist contain remnant crinoid ossicles, stems of sea-lilies, that have withstood the intense strain of rocks subducted to around 100 km depth. Spectacular S-C fabrics can be seen in all structural levels related to ductile shearing during exhumation (top-north, base-to-south). The 'extensional'

Fig. 11.31 S-C shear fabrics in calc-schists above the As Sifah eclogites, showing top-to-north fabrics

fabrics are related to uplift of footwall eclogites during exhumation back along the same subduction zone channel. The earlier prograde burial fabrics have mainly been overprinted by the later retrograde phase of deformation.

A dirt track runs north inland and the main beach north of As Sifah village is now one of the few public beaches remaining. Hence it can become crowded at weekends with campers and picnickers. Environmental destruction is becoming a major issue, especially with litter. One major development has already been built along the main As Sifah beach to the south with three new hotels, a marina, golf course and villas. It is imperative that this development does not extend to the north of As Sifah village and to the eclogite beach beyond. There are plans for further developments south of As Sifah village. The As Sifah eclogite beach, 200 m north of the main As Sifah beach, is completely surrounded by cliffs and accessible only by hiking around the headland to the north of the main Sifah beach. Along the coast to the north another larger beach is surrounded by cliffs of calc-schists with smaller eclogite boudins, structurally above the main eclogite (Fig. 11.33). Further north towards As Sheik inlet and village, the coast becomes increasingly more inaccessible and can only be approached from the sea.

Fig. 11.32 Small-scale sheath fold in As Sifah eclogites

Fig. 11.33 Photograph from a drone looking west at the eclogite beach, As Sifah

References

Glennie KW (2005) The deserts of Southeast Arabia. GeoArabia. Manama, Bahrain, 215 p

Miller JM, Gray DR, Gregory RT (2002) Geometry and significance of internal windows and regional isoclinal folds in northeast Saih Hatat, Sultanate of Oman. J Struct Geol 24:359–386

Nolan SC, Skelton PW, Clissold BP, Smewing JD (1990) Maastrichtian to early Tertiary stratigraphy and paleogeography of the central and Northern Oman Mountains. In: Robertson AFH, Searle MP, Ries AC (eds) The geology and tectonics of the Oman region. Geological Society, London, Special Publication no. 49, pp 495–520

Salm R, Baldwin R (1991) Snorkelling and diving in Oman. Motivate Publishers, Dubai (reprinted 2007)

Searle MP, Waters DJ, Martin HN, Rex DC (1994) Structure and metamorphism of blueschist–eclogite facies rocks from the northeastern Oman Mountains. J Geol Soc Lond 151:555–576

Searle MP, Warren CJ, Waters DJ, Parrish RR (2003) Subduction zone polarity in the Oman Mountains: implications for ophiolite emplacement. In: Dilek Y, Robinson PT (eds) Ophiolites in earth history. Geological Society, London, Special Publication no. 218, pp 467–480

Searle MP, Warren CJ, Waters DJ, Parrish RR (2004) Structural evolution, metamorphism and restoration of the Arabian continental margin, Saih Hatat region, Oman Mountains. J Struct Geol 26:451–473

Warren CJ, Parrish RR, Waters DJ, Searle MP (2005) Dating the geologic history of Oman's Semail ophiolite: insights from U-Pb geochronology. Contributions to Mineralogy and Petrology. https://doi.org/10.1007/s00410-005-0028-5

Saih Hatat

12

The Saih Hatat culmination is an extraordinarily large anticlinal fold structure formed during the later stages of ophiolite obduction in the Latest Cretaceous and then enhanced during the post-obduction phase, probably during the Oligocene–Miocene, or even younger (Fig. 12.1). Along the northern part of Saih Hatat major ductile shear zones striking roughly east-west and dipping north formed during exhumation of the footwall high-pressure eclogite and blueschist facies rocks (Fig. 12.2). At least four major ductile shear zones successively place lower pressure rocks above higher pressure rocks (from base: As Sifah, As Sheik, Hulw and Yenkit shear zones), with the higher structural levels all in carpholite or blueschist facies (Ruwi mélange) beneath the Muscat–Muttrah ophiolite. The World's largest sheath fold is beautifully exposed in the northern Saih Hatat region along Wadi Mayh. The ductile shear zones and top-north ductile S-C fabrics reflect uplift and exhumation of deeper thrust sheets beneath 'passive roof faults'. A major phase of uplift occurred after deposition of the Paleocene–Eocene limestones when normal faulting occurred around its perimeter reflecting uplift of the footwall shelf carbonates and basement relative to the overlying Late Cretaceous thrust sheets (Semail ophiolite, Haybi and Hawasina complexes).

The Saih Hatat culmination exposes extensive outcrops of Precambrian low-grade schists (Hatat schists), dolomites (Hijam Formation) and Ordovician Amdeh Formation sandstones and quartzites, in the core of the anticlinal structure as well as along the complex folds in the north. Middle Permian to Cretaceous shelf carbonates unconformably overlying basement rocks occur all the way around the dome with flat-lying undeformed sequences along the southern margin of Saih Hatat. However, along the northern margin these rocks become highly folded and sheared and progressively more affected by increasing high-pressure metamorphism towards the northeast at progressively deeper structural levels. The deepest levels are exposed around the coastal village of As Sifah with eclogites

© Springer Nature Switzerland AG 2019
M. Searle, *Geology of the Oman Mountains, Eastern Arabia*,
GeoGuide, https://doi.org/10.1007/978-3-030-18453-7_12

Fig. 12.1 Landsat photograph of the southern part of Saih Hatat region and the Ibra ophiolite block, showing main locations

formed at pressures in excess of 20 kbar and depths around 100 km in a NE-dipping Late Cretaceous subduction zone. A composite cross-section of the northern part of Saih Hatat is shown in Fig. 12.3. with a restoration of the Late Cretaceous thrusts and the High-pressure rocks in Fig. 12.4. A large scale NNE aligned sheath fold is spectacularly exposed along Wadi Mayh with a subsidiary parallel sheath fold beautifully exposed in the mountains northeast of Wadi Aday. The flat-lying synclinal sheath folds close to the SSW but faces (youngs) towards the NNE. The youngest rocks in the fold are Permian and the older Amdeh and Precambrian Hijam Formation dolomites wrap all around the fold. The uppermost anticline is the Jebel Qirmadil fold exposed in the far northern part of Saih Hatat.

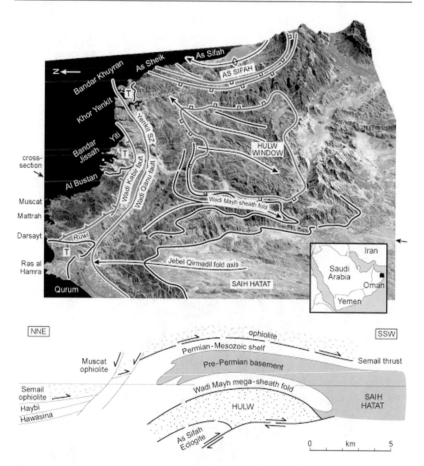

Fig. 12.2 Digital elevation model of the Saih Hatat culmination, view towards SE, showing the main geological features and cross-section illustrating the major fold nappe structures

South of the Saih Hatat anticline a large area of ophiolite rocks (Semail and Wadi Tayyin massifs) crops out east of the Semail Gap and south of Saih Hatat. The almost complete ophiolite section is seen around the Semail and Maqsad areas where mapping has shown the presence of a large mantle diapir structure (Nicolas et al. 1988). Some important outcrops of sub-ophiolite metamorphic sole occurs along the base of the ophiolite mantle sequence along Wadi Tayyin, and further

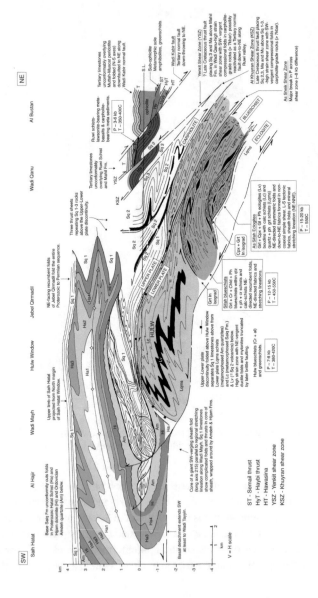

Fig. 12.3 Composite structural cross-section of northern Saih Hatat showing the major fold-nappes, overlying the Hulw dome and deepest As Sifah eclogite unit

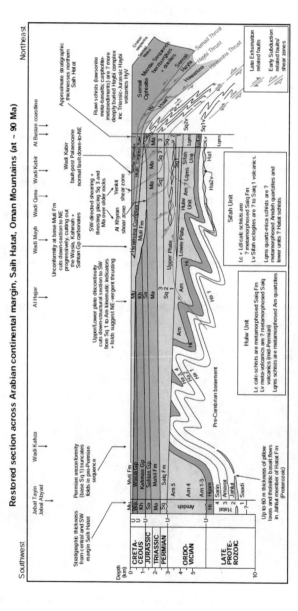

Fig. 12.4 Restoration of the shelf carbonates and Muscat peridotites shown in Fig. 12.3. To position prior to the subduction of the continental margin and high-pressure eclogite facies metamorphism

west towards Fanjah. The margin of the shelf carbonates here is also a late-stage steep, south-dipping normal fault, reflecting late-stage uplift of the Saih Hatat culmination.

12.1 Semail Ophiolite

One of the Semail ophiolite 'type examples' is the large section centered around Semail village toward Ibra, and the area to the east linking with the Wadi Tayyin ophiolite 'block' (Fig. 12.5). These ophiolite sections were originally mapped in detail by Bob Coleman, Cliff Hopson, Ed Bailey and their USGS colleagues in the 1970s (Coleman and Hopson 1981; Hopson et al. 1981; Pallister and Hopson 1981; Gregory 1984), and have more recently been mapped in detail by Nicolas et al. (1988) and Boudier et al. (1988). Both groups interpreted the Semail ophiolite sequence in this region as having formed along a Mid-Ocean ridge during the Late Cretaceous. A key factor in this interpretation is that, in this region, the only pillow lavas exposed are the earliest, stratigraphically lowest V1 series MORB-type lavas

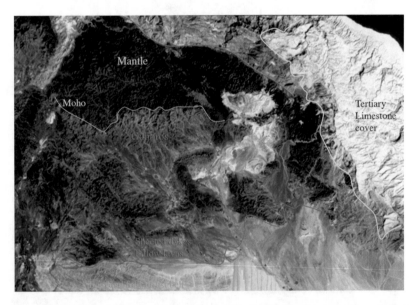

Fig. 12.5 Landsat satellite photograph of the Ibra block ophiolite showing main geological features

(Geotimes unit). The upper boninitic and island arc lavas discovered by the Open University group (Pearce et al. 1981; Alabaster et al. 1982; Lippard et al. 1986) in northern Oman (Chap. 8) are missing along the top of the ophiolite in the Ibra area. The V2 (Lasail) arc lavas were instrumental in defining the supra-subduction zone origin of the Semail ophiolite (Pearce et al. 1981).

The main road from the Semail Gap to Ibra and Sur provides some excellent exposures of the ophiolite complex. At the village of Luzuq, north of Semail, sheeted dykes are present in the axis of a synform where partially melted hornblende gabbros have pale-coloured leucosomes of plagiogranite. It is these felsic plagiogranites composed of plagioclase (albite to labradorite) with minor amounts of quartz, hornblende, and Fe–Ti oxides that contain minor amounts of zircon that can be used for U-Pb dating of ophiolite crustal sequences. South of the village of Dasir nearly vertical two-pyroxene layered gabbros are occasionally overturned dipping south towards the Moho. Alignment of plagioclase crystals in the gabbro clearly defines a magmatic lamination. The base of the crust is marked by layered gabbros with magmatic fabrics, whereas the higher gabbro levels can be seen to feed sheeted dykes structurally above. Dark weathering clinopyroxene-bearing wehrlite intrusions cross-cut the gabbros in this locality. In the northern Oman ophiolite these wehrlites are more common, for example around the Fanjah area, and thought to be associated with the later subduction-related V2 or Lasail volcanic rocks.

To the east of the Semail Gap, a large area of mountains is composed entirely of harzburgite with minor amounts of dunite (Fig. 12.6). Nicolas et al. (1988) proposed the existence of a large mantle diapir (Maqsad diapir) on the basis of mapping foliations and lineations defined by olivine and pyroxene crystals. Mapping of mantle fabrics, defined by foliations and lineations in harzburgites show a dramatic steepening in the Maqsad area, associated with a thick and complex Moho Transition zone. The MTZ in this area is over 400 m thick, shows thick gabbroic lenses, 100 m thickness of compacted dunites, and a cluster of chromite deposits (Nicolas et al. 2011). These authors proposed that the thickness of the MTZ is 0 at the ridge axis, thickening to about 100 m off-axis to over 400 m above the mantle diapirs. The MTZ is able to store a large volume of melt delivered from the mantle in the gabbro lens, which is then fed upwards into the oceanic crust. Magmatic fabrics are evident in the lower gabbros, interpreted as the magma chamber. A network of clinopyroxene, pargasite-bearing gabbroic dykes are thought to be the result of high-temperature (~ 700 °C) hydrothermal alteration

Fig. 12.6 Old photograph (ca. 1975) from Jebel Nakhl, view east over the Semail Gap to the harzburgite mountains of the Semail ophiolite in the Ibra block

close to the oceanic ridge axis. These dykes grade into lower temperature (500–400 °C) veins associated with seawater contamination.

It is likely that the Semail and Wadi Tayyin ophiolite sheets were once continuous across the Saih Hatat anticline with the ophiolite complex exposed around the Darsayt–Muscat–Muttrah area in the north. This would imply the obduction of a single giant thrust sheet of ophiolite all the way across the Oman shelf carbonate platform during the obduction event. The ophiolites along both sides of the Saih Hatat dome have a full ophiolite sequence of upper mantle and oceanic crust. The Muscat ophiolite seems to lack the characteristic metamorphic sole, although it is possible that this has been faulted out during subsequent exhumation of the high-pressure rocks. However, it could also be possible that the Muscat–Muttrah ophiolite might have been continuous with Gulf of Oman crust and upper mantle offshore the north coast of Oman. Bathymetry offshore the Muscat–As Sifah area shows a rapid deepening to depths of around 3 km some 30 km offshore, with large canyons carved into the continental shelf.

12.2 Wadi Tayyin Metamorphic Sole

The 'Green Pool' locality in Wadi Tayyin is another very important geological site where amphibolites and greenschists of the ophiolite metamorphic sole are well exposed (Ghent and Stout 1981; Hacker et al. 1996; Gray et al. 2000; Cowan et al. 2014). Like the Sumeini Window locality (Chap. 8), the metamorphic isograds and facies are inverted showing that the heat must have been derived from the hot peridotites of the overlying mantle wedge. U-Pb zircon and $^{40}Ar/^{39}Ar$ ages from amphibolites show that metamorphism was almost synchronous with ophiolite formation (Hacker 1993; Warren et al. 2005; Rioux et al. 2016). Together with the Masafi locality in UAE and the Sumeini Window in Northern Oman this site is one of the most important sites for metamorphic sole rocks that can be used for interpreting the obduction history of the ophiolite.

At Green Pool the metamorphic sole consists of 230 m thickness of strongly sheared hornblende + plagioclase amphibolites (Figs. 12.7 and 12.8). Garnet with abundant clinopyroxene inclusions occurs in the uppermost 10 m of the amphibolite sole, immediately below the peridotite contact (Figs. 12.9). Small knots of garnet + clinopyroxene (diopside) are granulite facies assemblages, but are contained within hornblende + plagioclase amphibolites, sometimes with garnet and occasionally with diposide (Fig. 12.10). Clinopyroxene occurs up to 80 m beneath the peridotite contact. Partial melt pods containing plagioclase + quartz + hornblende ± clinopyroxene occur within the uppermost few meters of the sole. A prominent band of quartzite (meta-chert) occurs towards the base of the amphibolites. All the sole rocks have been affected by later imbrication and folding (Fig. 12.11). Fabrics indicate a major component of pure shear flattening together with top–SW simple shear. Thermobarometry indicates formation of the peak granulite facies rocks at 770–900 °C and 11–13 kbar (Cowan et al. 2014). Pressures are much higher than can be accounted for by the thickness of the preserved ophiolite (15–20 km) leading to the model of NE-directed subduction of basaltic and minor chert-rich sediments to depths of >45 km beneath the Semail ophiolite during its formation (Searle 2007; Cowan et al. 2014). The age of the metamorphic sole is derived from U-Pb dating of zircons extracted from the granulite and amphibolite sole rocks, especially the small partial melt pods exposed near the top of the sequence. The age of peak metamorphism in sole rocks along the mountain range is between 96.17 and 96.14 million years (Rioux et al. 2016), timing that

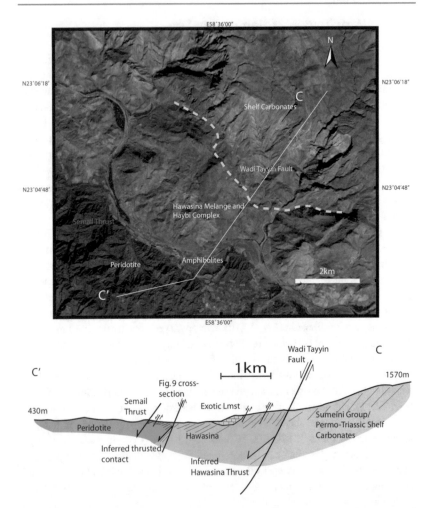

Fig. 12.7 GeoEye (Google Earth) satellite image of the Wadi Tayyin area, and cross-section showing the large down-south normal fault that bounds the southern margin of the Saih Hatat shelf carbonates, after Cowan et al. (2014)

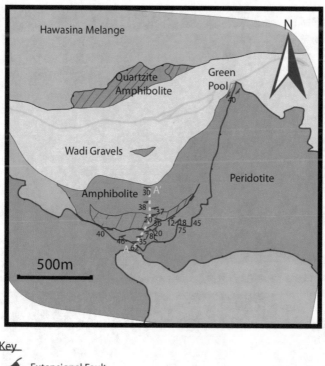

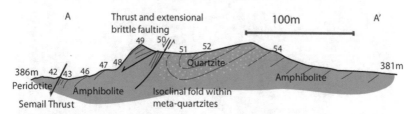

Fig. 12.8 Simplified geological map and cross-section of the Green Pool area, Wadi Tayyin metamorphic sole locality, after Cowan et al. (2014). Harzburgites of the mantle sequence have been thrust over garnet + diopside amphibolites of the metamorphic sole at the Green Pool locality

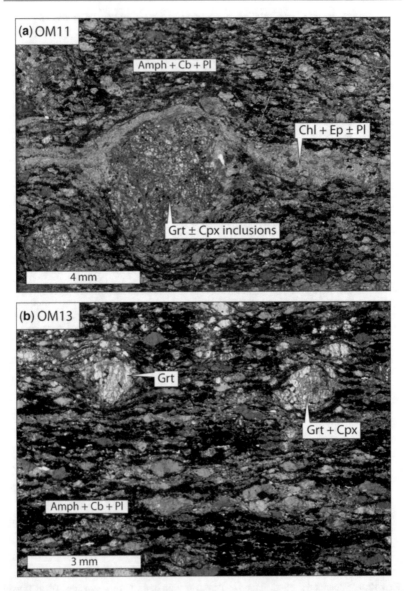

Fig. 12.9 Photomicrographs of metamorphic sole amphibolites from the Green Pool locality, Wadi Tayyin (plane polarised light). **a** Garnet porphyroblast with clinopyroxene inclusions wrapped by matrix foliation. **b** Garnet and clinopyroxene within a strongly foliated matrix comprising brown amphibole, plagioclase and carbonate

Fig. 12.10 Semail thrust with harzburgites of the mantle sequence (*left*) emplaced over garnet amphibolites of the metamorphic sole (*right*), Green Pool, Wadi Tayyin

overlaps with that of formation of the ophiolite crust. These data require that the ophiolite formed at the same time as basalts and cherts on the down-going plate were subducted to depths of more than 45 km beneath the ophiolite in a supra-subduction zone tectonic setting.

12.3 Wadi Kabir Fault

Along the northern Saih Hatat region the structures are dominated by north-dipping extensional faults and extensional S-C fabrics superimposed onto earlier S- or SW-vergent folds and cleavage. These shear zones and faults all relate to SW-directed thrusting of footwall high-pressure rocks beneath static 'stretching faults' where the motion all occurs along the footwall in a compressional tectonic environment. The structurally highest of these faults is the Wadi Kabir ('big wadi' in Arabic) fault that bounds the southern limit of the Muscat–Muttrah ophiolite and

Fig. 12.11 Green Pool locality in the metamorphic sole along Wadi Tayyin. **a** Isoclinal fold closure in meta-quartzites within the amphibolites of the sole. **b** Thrust faulted contact between the amphibolites and meta-quartzite unit showing discordant fabrics above and below

runs eastwards along Wadi Kabir towards Bandar Jissah. Late Paleocene–early Eocene shallow marine limestones overlie the Muscat peridotite along a planar unconformity that has dropped down to the north by about 100 m during post-Eocene faulting along the Wadi Kabir fault. The fault extends eastward separating Jurassic shelf carbonates along the footwall to the south from Paleogene limestones around Bandar Jissah to the north. The entire Cretaceous top of the shelf carbonates, allochthonous Hawasina and Haybi complex thrust sheets, and the ophiolite is missing along this fault.

The Wadi Kabir fault has seen at least three periods of motion along it: (1) Cenomanian-Turonian SW-directed thrusting of ophiolite peridotites over Haybi complex–Ruwi mélange, and Ruwi mélange over platform carbonates, (2) Santonian-Campanian normal sense shearing placing rocks of the Ruwi unit above Yiti unit rocks with Fe–Mg carpholite + pyrophyllite + sudoite metamorphic assemblages at pressures of about 8–10 kbar, and (3) post-Eocene top-north normal faulting related to uplift of footwall high-pressure rocks.

12.4 Ruwi Mélange

The Ruwi mélange underlies the Muscat–Muttrah ophiolite and is exposed all around the Ruwi valley and along the northern margin of Saih Hatat. It is a dominantly mudstone- or shaley matrix mélange enclosing blocks of shelf facies carbonates, and some lawsonite-bearing metabasalts. Pressure-Temperature conditions of formation are either 3–6 kbar and 250–315 °C (El-Shazli et al. 1994) based on lawsonite in meta-basalt, or pressures up to 10–11 kbar, and temperatures as low as 180 °C based on carpholite in the mudstone matrix. Carpholite-bearing meta-sediments occur across the whole of region from the Ruwi valley eastwards to Quriat, and in all structural units beneath the Muscat ophiolite and above the lower As Sifah eclogite units (Goffé et al. 1988; Agard et al. 2010). The carpholite group of minerals are single chain silicates that occur in Al-rich, low-temperature and high-pressure rocks. Assemblages consisting of Fe–Mg carpholite + pyrophyllite + chloritoid ± sudoite ± phengite mica give pressures of 4–5 kbar and temperatures of 340–280 °C and Fe-rich carpholite + kaolinite + quartz + illite + chlorite ± paragonite assemblages give PT conditions of 250–200 °C and 8–10 kbar (Goffé et al. 1988).

One large block of Lower Cretaceous shelf carbonate showing a spectacular recumbent isoclinal fold is exposed in the hills northwest of Ruwi (see Chap. 11). This is either an isolated block enclosed in the Ruwi mélange or a northern extent of the isoclinally folded shelf carbonates exposed south of the Ruwi valley.

Fig. 12.12 a View of Al Hamriyah district of Ruwi valley looking east. The Ruwi high-pressure unit is bounded by the Yenkit shear zone to the left and the Wadi Qanu fault to the right. **b** Exhumation-related top-to-north shear criteria (S-C fabrics) below Yenkit shear zone. **c** Carpholite-bearing quartz-rich boudin preserved within the shear bands, location shown on **b**. **d** Fresh carpholite hand specimen from Al Hamriyah, location shown on **a**, after Agard et al. (2010)

Top-north S-C deformation fabrics and large carpholite crystals between shear planes can be observed in the road cut above Al Hamriyah at the southern end of the Ruwi valley (Fig. 12.12). The Ruwi unit is separated from the overlying Yenkit–Yiti unit by the north-dipping ductile shear zone called the Yenkit shear zone (Searle et al. 2004; Agard et al. 2010). The Wadi Qanu fault is a later brittle fault cutting through the ductile shear zone and probably continuous with the large-scale normal fault that encircles the Saih Hatat dome.

12.5   Wadi Qanu Fault and Yenkit Shear Zone

The Wadi Qanu fault is a brittle thrust fault that cuts through the ductile Yenkit shear zone, and has been reactivated as a top-north normal fault during exhumation of footwall high-pressure rocks. Carpholite + pyrophyllite assemblages within shear zone rocks have pressures between 4 and 7 kbar and temperatures less than 300 °C indicating rapid burial of mudstones and limestones to depths of about 25 km followed by exhumation along these shear zones during the Late Creta-ceous. Along Wadi Qanu, Saiq 3 Formation limestones have been thrust south over younger Mahil Formation limestones. Early south-vergent duplex formation can be seen in the cliffs above Yiti village (Fig. 12.13). Later extensional fabrics indicate top-north, corresponding to footwall rocks being exhumed southwards. These NNE 'extensional' fabrics were superimposed on the earlier SSW thrust structures. Towards the west this fault curves into alignment with the Saih Hatat normal fault that encircles the shelf carbonates around the Saih Hatat dome.

A major ductile shear zone approximately 50 m thick termed the Yenkit shear zone cuts across from the Ruwi valley in the west to the head of Yenkit fjord then east to Bandar Khuyran and out to sea. It is a zone of intense shear fabrics, with isoclinal folds and strong cleavage separating Permian Formation limestones of the shelf carbonates below from Triassic Mahil Formation dolomites and limestones above. Although there appears to be little stratigraphic offset, the strain within the shear zone is high and the footwall rocks are more ductilely deformed than the hangingwall rocks. A large recumbent fold in the Triassic limestones above the shear zone can be seen to the west of Khawr Yenkit. Carpholite assemblages are found along the footwall rocks and within the ductile shear zone at Al Hamriyah in the Ruwi valley.

The Wadi Qanu fault–Yenkit shear zone has had multiple episodes of motion including: (1) NNE-facing folding related to NNE-directed subduction of the shelf carbonates to depths of about 20 km during carpholite grade metamorphism, (2) SW-vergent folding and thrusting related to exhumation of footwall shelf carbonates following peak metamorphism, (3) late-stage top-north normal faulting,

North   South

Fig. 12.13 Yiti duplex showing early stages of a thrust duplex forming in Mahil Formation dolomites, Yiti village, Wadi Mayh

related to culmination of the Saih Hatat dome, and (4) a post-Eocene phase of compression when Cenozic cover rocks around Bandar Jissah were folded around NNW-SSE axes.

The Permian Saiq Formation and Triassic Mahil Formation limestones south of the Yenkit shear zone show several more ductile shear zones that show intense isoclinal folding, sheath folding and high strain. These rocks are all at high-pressure with carpholite and occasional chloritoid in the meta-sediments and crossite in meta-basalts. The structurally deeper Hulw unit exposed in a window east of Wadi Mayh is bounded above by a major ductile shear zone, termed the 'Upper plate–Lower plate discontinuity' by Gregory et al. (1998) and Miller et al. (2002). Agard et al. (2010) showed that pressure-temperature conditions of the rocks above and below were similar, revealing that the shear zone was a mainly horizontal detachment, with little or no vertical offset, at blueschist facies depths. It appears that all these high-level structural units were metamorphosed at similar pressures and depths (6–9 kbar; around 20–30 km depth), and were the result of stacking up of buoyant carbonate-rich thrust units that were unsubductable. Only the higher pressure and deeper As Sifah eclogite unit was subducted to far deeper levels (20–23 kbar; around 80–100 km depth) within the NE-dipping Late Cretaceous subduction zone.

12.6   Wadi Mayh Sheath Fold

The Wadi Mayh mega-sheath fold is the largest and by far the best exposed of any sheath fold in the World (Searle 2007; Searle and Alsop 2007; Cornish and Searle 2017). The meandering course of the wadi north of Al Hajar village reveals magnificent 3-dimensional views across and along this sheath fold which shows a SSW elongation (x axis) of at least 15 km and possibly up to 25 km (Fig. 12.14). Sheath folds are highly curvilinear folds with axes elongated and rotated into the transport direction and this structure is aligned NNE-SSW parallel to the thrust direction of the Late Cretaceous nappes. Cornish and Searle (2017) mapped out the Wadi Mayh sheath fold in intricate detail using high-resolution photo panoramas and drone photography (Figs. 12.15 and 12.16). Magnificent exposures of the internal structure of the sheath fold can be seen along the winding course of Wadi Mayh. At one particularly impressive locality three 'eye' closure folds can be seen stacked on top of one another in a cross section view across the structure. Brown-weathering sandstones and quartzites of the Amdeh Formation wrap around the entire sheath fold which can also be seen to close at the southern end of the Wadi Mayh gorge (Fig. 12.17). Internal ductile shear zones, boudins, flattening fabrics and strong NNE-aligned lineations can all be clearly seen in water-worn wadi walls (Figs. 12.18 and 12.19). The rocks to the north of the sheath fold are all in high-pressure carpholite facies ($\sim$9–11 kbar; 25–30 km depth of burial; <330 °C; Agard et al. 2010) a subduction-related high-pressure terrane that eventually extends to the deepest rocks exposed in Oman, the As Sifah eclogites (20–23 kbars, 80–100 km depth of burial).

The entire profile along Wadi Mayh to Yiti and east towards As Sifah is unique in any sub-ophiolite complex in the World. The rocks record progressive burial and exhumation of the thinned and stretched leading margin of the Arabian continental crust down a subduction zone to depths up to 80–100 km. The narrow Wadi Mayh gorge through the sheath fold is very prone to flash flooding during winter rainstorms and the road is frequently washed out. During Cyclone Gonu, the Wadi Mayh gorge filled up to a depth of over 40 m. During any future rebuilding work on the Wadi Mayh road it is imperative that the mountains along the flanks of the

(a)

(b)

Fig. 12.14 Wadi Mayh sheath fold looking SSW showing the eyes of the main sheath fold picked out by bedding (yellow) in Saiq Formation limestones capped and rimmed by the brown-weathering Amdeh Formation sandstones around the sheath fold

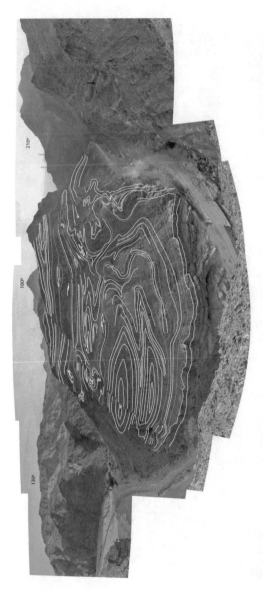

Fig. 12.15 Panorama (WM5) of the Wadi Mayh sheath fold in Permian Saiq Formation limestones, from Cornish and Searle (2017). Two distinct sheaths are stacked atop one another but are wrapped around by continuous beds

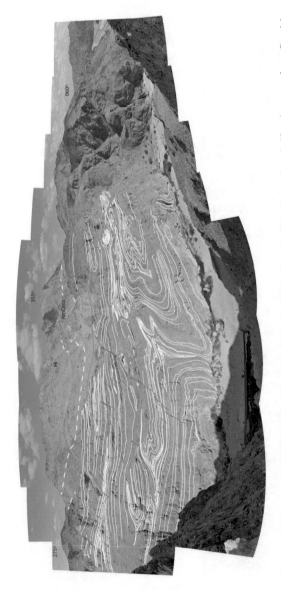

Fig. 12.16 Panorama (WM20) of the Wadi Mayh sheath fold in Permian Saiq Formation shelf carbonates, from Cornish and Searle (2017). Structures exposed along the west side of the Wadi Mayh gorge, showing the same cascading fold limb wrapping around the same cat's eye exposures in WM5

Fig. 12.17 Panorama (WM27) of the Wadi Mayh sheath fold in Permian Saiq Formation shelf carbonates, from Cornish and Searle (2017). View west along the southern limit of the canyon showing the sheath fold syncline hinge plunging down to the south, with brown Amdeh Formation sandstones wrapping around the hinge

Fig. 12.18 Spectacular folds in Permian Saiq Formation shelf carbonates along the western wall of Wadi Mayh canyon

wadi are preserved intact to preserve this unique and important geological site. There are other easier alternative access sites for a new road to Yiti to the west or east of the Wadi Mayh canyon, and this would preserve the magnificent wadi outcrops along the sheath fold.

12.7   Wadi Aday Fold Nappes

Another excellent profile across the folds and thrusts of northern Saih Hatat is exposed along Wadi Aday, especially the profile from Al Amerat village in Saih Hatat north to the Ruwi valley. The stratigraphy of the northern Saih Hatat region is illustrated in Fig. 12.20. Beneath the mid-Permian unconformity Precambrian (Hatat schists) and Palaeozoic rocks are exposed in the Saih Hatat window. The Hijam dolomite, Ordovician Amdeh sandstones and Lower Permian Al Khlata glacial sediments are each separated by unconformities. The Middle and Upper

Fig. 12.19 Folds and detachments exposed along the Wadi Mayh sheath folds in the central part of the canyon

Permian Saiq Formation shelf carbonates onlap all previous units above the highest unconformity are approximately 650 m thick and have been divided into five main units (LeMétour et al. 1990). A lower thin band of conglomerate with dolomite clasts is overlain by a prominent band of pale-coloured volcanic rocks, mainly tuffs (Sq1) that are a good marker band for tracing around the fold nappes. Above are massive thick-bedded dolomites (Sq2) with two interbanded layers of tuffs and pillow lavas (Sq3). Approximately 150 m thickness of well-bedded grey micritic dolomite and limestone (Sq4) is overlain by dolomites with stromatolites and spary limestones with thin yellow shale bands (Sq5).

The structure of the Wadi Aday profile is dominated by a large-scale flat-lying NNE facing sheath fold with an upper recumbent anticline (Jebel Qirmadil fold) and underlying isoclinal syncline that has been sheathed towards the SSW (Cemetery syncline). This flat-lying syncline axis can be traced though the mountains northwest of Wadi Aday, and the 'eye' of the sheath fold is beautifully exposed above the Cemetery in the northwestern corner of Saih Hatat. The Wadi Aday sheath fold axis is parallel to the Wadi Mayh sheath fold axis 10 km to the SE. All of these folds are relatively flat-lying; hence the axial traces follow

Upper Plate Stratigraphy

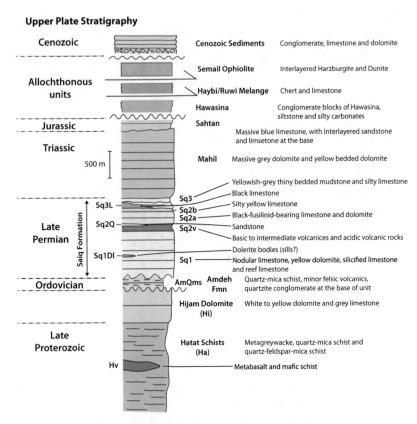

Fig. 12.20 Stratigraphic section of rocks preserved in the northern part of the Saih Hatat Window from the Precambrian basement units up to the Semail ophiolite

contours around the mountains. All folds in this area are NNE-facing (i.e. stratigraphy 'younging' towards the NNE), similar to the Wadi Mayh sheath fold.

Along Wadi Aday the basement units (Hatat schists, Hijam dolomite, Amdeh sandstones) and their intervening unconformities are infolded with the recumbent fold nappes (Fig. 12.21). A prominent conglomerate unit east of Wadi Aday may be a part of the Lower Permian Al Khlata Formation. A major NNE-verging thrust, the Jebel Qirmadil thrust, underlies the fold-nappe and cuts up-stratigraphic section towards the northwest. This thrust places the Jebel Qirmadil–Cemetery isoclinal to recumbent fold pair above the less deformed Triassic Mahil Formation dolomites.

Fig. 12.21 View looking NW across Wadi Aday showing complex fold patterns in the Wadi Aday sheath fold

The thrust is spectacularly exposed in the mountains to the west of Wadi Aday. Above the thrust Permian limestones are intensely folded with flat-lying axial planes (Figs. 12.22 and 12.23). Below the thrust, well-bedded dark coloured Mahil Formation limestones and dolomites show regular east-west strike and dip at about 45° NNE beneath the Ruwi valley. Isoclinal folds in the limestones have stretched out axis where the outermost beds elongate into pointed structures termed 'narwhal' structures (Fig. 12.23).

The northern end of Wadi Aday is complicated with well-bedded Mahil Formation limestones dipping north beneath the Ruwi valley. A thin band of serpentinite lies along a major thrust plane west of Hatat House. The Ruwi mélange lies above this fault and occupies most of the Ruwi valley. Early thrusts have been reactivated as later normal faults during uplift and exhumation of the Saih Hatat dome. The largest of these faults is the Saih Hatat normal fault that encircles the shelf carbonates around the northern margin of the dome.

Fig. 12.22 Isoclinal folds in Permain Saiq Formation limestones, above Wadi Aday. *Note the stretched out axial traces of folds into 'narwhal' structures*

12.8   Jebel Qirmadil

Jebel Qirmadil above the northwestern corner of Saih Hatat marks the gradation from undisturbed shelf carbonates all around the southwest, and southern flanks of the dome to the intensely folded and thrust northern margin. From Saih Hatat the flat-lying mid-Permian unconformity can be traced northwards as far as Jebel Qirmadil where the first recumbent folds begin. The Jebel Qirmadil anticline axis is paired with the structurally deeper Cemetery syncline, which forms the main sheath fold in this part of the mountains. It is debated whether the Wadi Aday sheath fold is structurally the same fold as the larger Wadi Mayh sheath fold to the east, or a separate structurally higher sheath fold, secondary to the major Wadi Mayh sheath fold.

A climb of Jebel Qirmadil to the summit provides an excellent view-point in all directions. To the southwest the entire Saih Hatat core region can be seen with the

Fig. 12.23 Narwhal structures, the extended fold axes of isoclinal folds in Permian Saiq Formation shelf carbonates, above Wadi Aday

flat-lying mid-Permian limestones of the Saiq Formation above the unconformity. These rocks become entrained in the recumbent folds of Jebel Qirmadil and the Cemetery fold along northern rim of the Saih Hatat bowl, both of which comprise the Wadi Aday sheath fold. To the east these fold-nappe structures can be seen in all their structural complexity. The core region of the sheath fold is seen in the high mountains east of Jebel Qirmadil and west of Wadi Aday, with bedding in the Permian Saiq Formation limestones completely enclosed in rugby ball 'eye' structures. To the north, the mountains drop down to the Batinah coast and Qurum region, looking north along the axis of the main Saih Hatat fold structure.

12.9   Wadi Daykah

This spectacular narrow, vertical sided wadi canyon cuts right through the shelf carbonates from the village of Hayl al-Ghaf south to the settlement of Ghubrat al-Tam where it connects with the Semail Ophiolite along Wadi Tayyin

Fig. 12.24 The canyon of Wadi Daykah, the 'Devils Gap'. Mountains are composed of the Permian and Triassic parts of the shelf carbonates that unconformably overlie Permain Al-Khlata Formation and Ordovician Amdeh Formation

(Figs. 12.24 and 12.25). Wadi Daykah was the only access route across the mountains between Sur and Muscat. An excellent description of an epic camel journey through the wadi, called the 'Devil's Gap' was given by Col. Samuel

Fig. 12.25 Steep sided mountains along the Wadi Daykah 'Devils Gap'

Miles (1884), reprinted in Ward (1987): "*Towering loftily, sheer and perpendicular above the narrow floor, the huge walls of rock give the appearance as if the mountain range had been suddenly split in twain from the base to the summit by some convulsion of nature, exhibiting a singular illustration of impressive grandeur*" (Miles 1884). The Wadi Daykah canyon has been cut by aeons of floodwaters thundering down the chasm, in the words of Miles "*to provide an outflow for the pent-up waters of the Tyin valley through a mountain range in the most singular specimen of earth sculpture I have seen in Arabia*". The lower part of the canyon has recently been dammed.

Wadi Daykah contains important outcrops of the Ordovician Amdeh Formation, that extend eastwards to Wadi Sarin and Wadi Salil near Quriat. A section of quartzite sandstones, shales and shell beds approximately 690 m thick was studied across Wadi Daykah, west of the village of Mazara (Heward and Penny 2014; Heward et al. 2016). These rocks contain bivalves, crinoids, nautiloids, tribolites and the trace fossils *Cruziana* and *Skolithus* ichnofacies. Many of the best outcrops were flooded with the filling of the Wadi Daykah water reservoir behind a newly constructed dam. Wadi Daykah also shows good exposures of the Early Permian Al Khlata glacial deposits comprising silty-sandy diamictites with dropstones, which are widespread and form an important oil reservoir in the subsurface of Central and South Oman. The Al Khlata outcrops in the mountains are overlain disconformably by the Saiq Formation limestones that form the steep cliffs and high mountains either side of Wadi Daykah. Permian and Mesozoic limestones show large scale folding related to the widespread deformation associated with the exhumation of the High-pressure eclogites and blueschists of the As Sifah unit.

12.10 Wadi Serin

The high mountains along the southeastern part of Saih Hatat are composed of Permian Saiq Formation limestones and overlying Triassic Mahil Formation limestones overlying the Precambrian Hatat schists, Hijam dolomite and Palaeozoic rocks of the basement (Fig. 12.26). These Permian-Triassic limestones are unaffected by the large-scale folds and thrusts seen along the northern margin of Saih Hatat. These remote high mountains and wadis form the Wadi Serin Arabian Tahr (*Arabitragus jayakari*) reserve covering over 2400 km^2, where these rare, endangered, mountain goats are still living. Special permits are required to access the reserve. There are less than 3000 Arabian tahr most of which occur in the reserve area, inhabiting steep cliffs and high peaks. The spread of domestic goats has contributed greatly to the demise of the tahr. Other rare animals in the Wadi

Fig. 12.26 Shelf carbonate mountains overlying Hatat schists, Wadi Serin area, south-western margin of Saih Hatat

Serin reserve include Blandford's fox (*Vulpes cana*), Arabian mountain gazelle (*Gazella arabica*), Arabian wolf (*Canis lupus arabs*), as well as numerous reptiles including the Oman saw-tailed viper (*Echis* sp), and over 400 species of plants, ten of them endemic and at least four on the endangered species Red List of Omani plants.

References

Agard P, Searle MP, Alsop GI, Dubacq B (2010) Crustal stacking and expulsion tectonics during continental subduction: P-T deformation constraints from Oman. Tectonics 29, TC5018. https://doi.org/10.1029/2010tc002669

Alabaster T, Pearce J, Malpas J (1982) The volcanic stratigraphy and petrogenesis of the Oman ophiolite complex. Contrib Miner Pet 81(3):168–183

Boudier F, Ceuleneer G, Nicolas A (1988) Shear zones, thrusts and related magmatism in the Oman ophiolite: initiation of thrusting on an oceanic ridge. Tectonophysics 151:275–296

Coleman RG, Hopson CA (1981) Introduction to the Oman Ophiolite special issue. J Geophys Res 86:2495–2496

Cornish SB, Searle MP (2017) 3D geometry and kinematic evolution of the Wadi Mayh sheath fold, Oman, using detailed mapping from high-resolution photography. J Struct Geol 101:26–42

Cowan RJ, Searle MP, Waters DJ (2014) Structure of the Metamorphic Sole to the Oman ophiolite, Sumeini Window and Wadi Tayyin: implications for ophiolite obduction processes. In: Rollinson HR, Searle MP, Abbasi IA, Al-Lazki A, Al-Kindy MH (eds) Tectonic evolution of the Oman Mountains. Geological Society, London, Special Publication no 392, pp 155–176

Ghent E, Stout M (1981) Metamorphism at the base of the Samail Ophiolite, southeastern Oman Mountains. J Geophys Res 86:2557–2571

Goffé B, Michard A, Kienast JR, LeMer O (1988) A case of obduction related high P low T metamorphism in upper crustal nappes, Arabian continental margin, Oman: P-T paths and kinematic interpretation. Tectonophysics 151:363–386

Gray DR, Gregory RT, Miller JM (2000) A new structural profile along the Muscat-Ibra-transect, Oman: implications for the emplacement of the Semail ophiolite. In: Dilek Y, Moores EM, Elthon D, Nicolas A (eds) Ophiolites and oceanic crust: new insights from field studies and Ocean Drilling Program. Geological Society of America Special Paper no 349, pp 513–523

Gregory RT, Gray DR, Miller JM (1998) Tectonics of the Arabian margin associated with the emplacement of the Oman Margin along the Ibra Transect: new evidence from NE Saih Hatat. Tectonics 17:657–670

Hacker BR, Mosenfelder JL, Gnos E (1996) Rapid emplacement of the Oman ophiolite: thermal geochronological constraints. Tectonics 15:1230–1247

Heward AP, Penny RA (2014) Al-Khlata glacial deposits in the Oman mountains and their implications. In: Rollinson H, Searle MP, Abbasi IA, Al-Lazki A, Al-Kindy MH (eds) Tectonic evolution of the Oman mountains. Geological Society, London, Special Publication no 392, pp 279–302

Heward AP, Booth GA, Fortey RA, Miller CG, Sansom IJ (2016) Darriwilian shallow-marine deposits from the Sultanate of Oman, a poorly known portion of the Arabian margin of Gondwana. Geol Mag 155(1):59–84

Hopson CA, Coleman RG, Gregory RT, Pallister JS, Bailey EH (1981) Geologic section through the Samail Ophioliteand associated rocks along a Muscat–Ibra transect, Southeastern Oman Mountains. J Geophys Res 86:2527–2544

LeMétour J, Rabu D, Tegyey M, Bechennec F, Beurrier M, Villey F (1990) Subduction and obduction: two stages in the Eo-Alpine tectonometamorphic evolution of the Oman mountains. In: Robertson AFH, Searle MP, Ries AC (eds) The geology and tectonics of the Oman region. Geological Society, London, Special Publication no. 49, pp 327–339

Lippard SJ, Shelton AW, Gass IG (1986) The ophiolite of northern Oman. Blackwell Scientific Publications, Oxford, p 178p

Miles S (1884) Journal of an Excursion in Oman, in Southeast Arabia

Miller JM, Gray DR, Gregory RT (2002) Geometry and significance of internal windows and regional isoclinal folds in northeast Saih Hatat, Sultanate of Oman. J Struct Geol 24:359–386

Nicolas A, Boudier F, Ceuleneer G (1988) Mantle flow patterns and magma chambers at oceanic ridges: evidence from the Oman ophiolite. Mar Geophys Res 9:293–310

Pallister JS (1981) Structure of the Sheeted Dyke Complex of the Samail ophiolite near Ibra, Oman. J Geophys Res 86:2661–2672

Pearce JA, Alabaster T, Shelton AW, Searle MP (1981) The Oman ophiolite as a Cretaceous arc-basin complex: evidence and implications. Phil Trans R Soc Lond A300:299–317

Rioux M, Garber J, Bauer A, Bowring S, Searle MP, Keleman P, Hacker B (2016) Synchronous formation of the metamorphic sole and igneous crust of the Semail ophiolite: new constraints on the tectonic evolution during ophiolite formation from high-precision U-Pb zircon geochronology. Earth Planet Sci Lett 451:185–195

Searle MP (2007) Structural geometry, style and timing of deformation in the Hawasina Window, Al Jabal al-Akhdar and Saih Hatat culminations, Oman Mountains. GeoArabia 12:93–124

Searle MP, Alsop GI (2007) Eye-to-eye with a mega-sheath fold: a case study from Wadi Mayh, northern Oman Mountains. Geology 35:1043–1046

Searle MP, Warren CJ, Waters DJ, Parrish RR (2004) Structural evolution, metamorphism and restoration of the Arabian continental margin, Saih Hatat region, Oman Mountains. J Struct Geol 26:451–473

Warren CJ, Parrish RR, Waters DJ, Searle MP (2005) Dating the geologic history of Oman's Semail ophiolite: insights from U-Pb geochronology. Contributions to Mineralogy and Petrology. https://doi.org/10.1007/s00410-005-0028-5

Ward P (1987) Travels in Oman. Oleander Press, 571 p

Sharkiyah, Eastern Oman Mountains

13

The Sharkiyah or Eastern Oman Mountains include the all the mountains north of the Wahiba Sands that extend from the Quriat region eastwards to Ras al Hadd, and the most easterly tip of the north coast (Fig. 13.1). The northern coast between Quryat and Sur has a rocky coastline with a few coral and pebble beaches but one beach near Fins is a magnificent stretch of white sand in between two rocky headlands. Illegal collection of sand for construction sites in Ash Shab and Tiwi villages has seriously affected the beach that also suffers from rubbish dumping by weekend trippers. This eastern coastal region used to be wild and empty with only a few small fishing villages at Dibab, Bimmah, Fins and Tiwi. Herds of wild mountain gazelle roamed the coastal plains and the occasional Arabian tahr was seen in the higher peaks. Mountain gazelle are now uncommon in this region except for the remoter parts of the high mountains, and Arabian tahr are restricted to the high peaks around the Wadi Serin reserve. The construction of a major highway linking Quriat to Sur with several large new bridges spanning Wadi Daykah, Wadi Ash Shab and Wadi Tiwi has now opened up the region for development. The entire coastline from Sur to Quryat has been staked out, ready for development and it is crucial that any development plans enable preservation of the Ash Shab gorge and also the Fins beach area. The inland mountains (Jebel Abiad, Jebel Bani Jabir) however remain in a wild state and there are some excellent trekking routes across these mountains.

The mountains rise to 2300 m altitude around the Salmah plateau, a large upland region of white limestone crags along Jebel Abiad and Jebel Bani Jabir (Fig. 13.2). These mountains are formed of Paleocene to Early Eocene Jafnayn or

© Springer Nature Switzerland AG 2019

M. Searle, *Geology of the Oman Mountains, Eastern Arabia*,

GeoGuide, https://doi.org/10.1007/978-3-030-18453-7_13

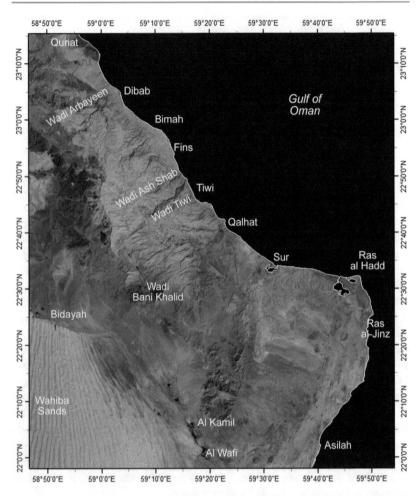

Fig. 13.1 Landsat photograph of the Sharkiyah (eastern) Oman Mountains showing key locations

Abat Formation and Middle and Late Eocene Seeb Formation limestones. The massif forms a large-scale box fold with steep flanks along the north and a gentle folded southern flank. One of the largest caves in the world, Majlis al-Jinn underlies the Salmah plateau, an area that undoubtedly has more caves to explore. Three large wadis drain the Salmah plateau, Wadi Arbayeen, Wadi Ash Shab and

Fig. 13.2 Landsat photograph of white limestone mountains of Jebel Bani Jabir and the Selma plateau. Dark-coloured rocks in the southwest are ophiolites and Haybi complex lavas and cherts

Wadi Tiwi. They have all carved deep gorges into the Paleocene-Eocene limestone and have perennial streams running through them with blue pools, cascading waterfalls and oases of oleander plants, ferns and palms lining the boulder strewn wadi floor.

Along the southern side of the mountains thick Paleocene–Eocene limestones form the high Salmah plateau and Jebel Bani Jabir both rising to around 2000 m. Spectacular canyons have been carved into these white mountains with hidden villages and stunning mountain scenery. Wadi Bani Kharus is one of the most beautiful wadis in Oman with large sections of flowing water and several deep blue pools surrounded by steep cliffs. The mountains drop down to the south to a gravel plain with the magnificent sandsea of the Wahiba Sands to the south. In places, the dunes of the Wahiba sands have swept right up to the mountains.

13.1   Wadi Arbayeen

Wadi Arbayeen inland from Dibab, is a beautiful scenic wadi that has a perennial river with waterfalls and magnificent blue pools ideal for swimming and exploring. A rough 4-wheel drive track ascends to the tiny village of As Suwayh, where one turning to the north connects through to Wadi Daykah, and the other dead-ends at a beautiful village nestling beneath huge cliffs of Paleocene limestones. One of the largest wadi pools with a waterfall provides excellent swimming opportunities (Fig. 13.3). The rocks along Wadi Arbayeen show Ordovician Amdeh quartzites and sandstones, overlain unconformably by Permian Saiq Formation shelf carbonates and truncated by the unconformity beneath the Paleocene limestones (Fig. 13.4). It is possible to climb up to the Selmah plateau from here although the route is circuitous and very steep. The views down Wadi Arbayeen towards Wadi Daykah and Quriat plains are magnificent.

Fig. 13.3 Wadi Arbayeen swimming pool and waterfall

Fig. 13.4 Old photograph (ca. 1984) of the track into Wadi Arbayeen

13.2   Dibab (Bimmah) Sink Hole

The coastal plain between Quriat and Fins is composed of gently north-dipping Eocene limestones (Seeb Formation) which are unconformably overlain by almost flat-lying lithified Pleistocene beach sands. The whole area is a classic karst with typical limestone weathering, riddled with caves and underground wadi systems that connect to the sea. Southeast of Quriat near the village of Dibab and about one kilometer inland is a large sinkhole 50–60 m wide and 25 m deep, formed when Eocene limestones dissolved and collapsed above a deep cave system (Fig. 13.5). The hole is accessible and filled with brackish water but the cave system linking it to the sea is inaccessible and beneath water level. The Bimmah sinkhole has now been developed as a tourist attraction with picnic areas and concrete steps leading down to the pool. It is a popular spot for high divers jumping off the cliffs into the blue waters of the cave. The Dibab sink hole is the largest of a series of such karstic

Fig. 13.5 Dibab (Bimmah) sink hole

features in the coastal plain between Quriat and Sur. Water draining off the Salmah plateau to the south flows underground along the Fins–Bimmah coastal plain with sea-water also penetrating in during high tides. The wadis here are prone to sudden flash floods during the winter rains.

13.3   Majlis al-Jinn Cave

The Salmah plateau reaching a height of 2200 m is riddled with caves and the largest of these, Majlis al-Jinn ranks amongst the top ten largest caves in the World approximately 310 m long, 225 m wide and is between 140 and 178 m high (Fig. 13.6). There are three small entrances and abseil (rappel) bolts have been placed to enable cavers to descend by a free-fall abseil of approximately 130 m. Exploration of the cave is a serious undertaking and a permit from the Ministry of Tourism is required. Although the abseil (rappel) descent into the great void is exhilarating, the ascent by jumaring can be extremely tiring. Only experienced climbers or cavers should attempt this. Majlis al-Jinn, together with several other deep caves on the Salmah plateau, was initially surveyed by Don Davison and Cheryl Jones in the late 1980s. The top of the cave is Middle Eocene Nummulitic limestones of the Seeb Formation. The Paleocene–Eocene limestones that make up the Salmah plateau are over 1200 m thick and folded into a giant box fold. Thin Maastrichtian conglomerates (Qahlah Formation) overlie eroded ophiolite rocks, mainly mantle sequence harzburgites that are strongly serpentinised.

Majlis al Jinn is reached by a steep 4-wheel drive track leaving the coastal road east of Bimmah village. The road ascends the Salmah plateau to the spectacular small village of Umq (Umq al Rubakh) with palm plantations and a large cave. Perennially flowing streams flow past the village and after rains there is a waterfall straight above the village. From Umq a trail leads up the jebel to the small holes that mark the top of the Majlis al Jinn. Another 4-wheel drive track winds up the Salmah plateau from near Fins village on the coast past the tiny villages of Hir-aymah and Ta'ab to the small hamlet of Salmah (Fig. 13.7). From here the Majlis al Jinn can be reached by a short walk. Ancient tombs, intricately build of numerous small flat stones, up to 10 m in height can be seen along the highest ridge near Majlis al-Jinn.

Fig. 13.6 Majlis al-Jinn cave on the Selmah plateau

Fig. 13.7 The Salmah plateau at around 1500–2000 m elevation

13.4   Fins Beach

Fins beach is one of the few white sand beaches along this stretch of the north-eastern Oman coastline, and for that reason is now very popular with day trippers and campers, especially during weekends and holidays (Fig. 13.8). The headland to the west of the beach is a great spot for ocean watching for sea-birds and fish life. Turtles, dolphins, rays and occasional whales can be spotted from the cliff-tops. The coast to the east of the beach has low Eocene limestone cliffs with some impressive overhangs eroded out by the sea and also from cliff-sapping molluscs. Inland of Fins beach four old, flat-lying terraces are clearly visible and provide evidence of recent uplift of this part of the Oman Mountains (Fig. 13.9). The oldest and highest terrace is 200 m above present sea-level, but appears to be fairly restricted to the Fins area, and cannot be traced far to the east or west. This evidence for uplift of the land around Fins and Tiwi is in stark contrast to the drowned coastlines around Bandar Khuyran, Bandar Jissah and the Muscat area to

Fig. 13.8 Coastline around Fins–Wadi Ash Shab

Fig. 13.9 Fins beach with four distinct marine terraces in the cliffs behind the beach

the west. At Fins the terraces are cut into Paleocene and Early Eocene limestones which show extensive karstic weathering and have numerous sink-holes. The terraces contain numerous molluscs and bivalves, but the precise ages are as yet unknown. Hoffman et al. (2016) describe large blocks of Eocene limestones up to 7 m by 1.5 m that they interpret as tsunami deposits. These authors speculate that the source was an earthquake along the Makran subduction zone in the Gulf of Oman.

13.5 Ash Shab Gorge

Wadi Ash Shab drains the northern part of the Salmah plateau, an uplifted dome comprising Paleocene-Eocene limestones that unconformably overlie the eroded ophiolite mantle sequence peridotites (Figs. 13.10 and 13.11). The limestones contain numerous fossils, including numerous species of corals, echinoids, bivalves, gastropods and the larger foraminifera such as *Nummulites*. The limestones have a karstic weathering with numerous caves and underground drainage channels. Wadi Ash Shab is a famously scenic beautiful wadi with perennial

Fig. 13.10 Old photograph (ca. 1985) of Wadi Ash Shab gorge

Fig. 13.11 Wadi Ash Shab

flowing water, numerous small waterfalls, and running streams linking wadi pools lined with flowering oleanders and palm trees (Fig. 13.12). A walking trail has now been constructed along the wadi to a stretch of magnificent deep blue pools.

Fig. 13.12 Wadi pools in the Ash Shab gorge

Further access along the wadi is only by swimming through pools and scrambling (Figs. 13.13 and 13.14). At the end a deep pool leads to a narrow underwater canyon with only a few inches of air space to progress. A short dive beneath the rock overhang and then one surfaces into a beautiful, magical cave with a small waterfall at the end. It is possible to climb alongside the waterfall (an old rope hangs down this section) and continue up the wadi to the next system of pools.

Fig. 13.13 Pools leading up to the blue grotto in Wadi Ash Shab

A higher trail bypasses the pools and heads higher up into the canyon. It is possible to link treks from Wadi Ash Shab across to the southern side of the mountains. A maze of wadis draining the southern slopes of this area mean that navigation is

Fig. 13.14 Swimming in Wadi Ash Shab pools

crucial to avoid descending the wrong wadi system. Two days are needed to trek across the Salmah plateau from Wadi Ash Shab or Wadi Tiwi across to Wadi Bani Khalid. The narrow deep canyon of Wadi Ash Shab is particularly prone to flash floods, when the wadi can fill up to a depth of over 20 m in places. Extensive erosion and damage occurred during the devastating floods accompanying Cyclone Gonu in 2007.

13.6 Wadi Tiwi

Wadi Tiwi, only 4 km east of Ash Shab is another large and beautiful wadi cutting deep into the eastern mountains (Fig. 13.15). The wadi also has perennial flowing streams, small waterfalls and many magnificent pools surrounded by palm trees, reed beds and vegetation. The beautiful old Omani mountain village of Mihbam, perched on a bluff overlooking the wadi, with numerous date and lemon trees and an ancient fort lies a few kilometres up the wadi (Fig. 13.16). It is possible to drive nearly 8 km to the end of a steep and tricky 4-wheel drive track. It is also possible to trek along well made trails up the cliffs to the southeast of the main wadi, and even across to Wadi Bani Khalid, a two-day hike across Jebel Bani Jabir, crossing a pass at 2200 m. The trek across the mountains is a long and meandering two-day route involving roughly 25 km of trekking, and requires good navigation skills, or hiring a local guide (Dale and Hadwin 2001). The floor of Wadi Tiwi has eroded

Fig. 13.15 Wadi Tiwi showing the upper eroded ophiolite mantle sequence overlain by thick Paleocene-Eocene limestones

Fig. 13.16 The village of Mihbam, Wadi Tiwi

Fig. 13.17 Blue pools and lush vegetation along Wadi Tiwi

down through Eocene and Paleocene limestones to the top of the ophiolite which is composed of highly serpentinised mantle sequence harzburgite. The entire crustal sequence of the ophiolite has been eroded off, prior to deposition of the overlying Paleocene limestones. The lower part of Wadi Tiwi has some beautiful blue pools lined with palms and reed-beds (Fig. 13.17).

13.7 Qalhat and Jebel Bani Jabir

Qalhat, 20 km west of Sur, is one of the most important archaeological sites in Oman and became a UNESCO World Heritage site in 2018. Qalhat was an important trading city in the ancient Kingdom of Hormuz, with fortified walls, streets and houses, the remains of which can still be seen. The most important ruin is the Bibi Maryam mosque and mausoleum which was built in 1312. Qalhat was visited by Marco Polo in the 13th century and Ibn Battuta in the 14th century and is recorded in both their writings. The town was also sacked by the Portugese and is reputed to have been affected by an earthquake on the Qalhat fault in the 15th century. A new tarmac road now runs along the coast connecting Quriat via the fishing villages of Bimmah, Fins, Tiwi and Qalhat to the major town and port of Sur. One of the most impressive mountain tracks for 4-wheel drive vehicles is a new track that zig-zags up the mountains from Qalhat, crosses the barren upland mountains of the Jebel Bani Jabir plateau and descends to the village of Ish-maiyyah in Wadi Khabbah, on the southwest side of the main mountain range. From Ishmaiyyah it is possible to drive south and then west to the town of Ibra. Another 4-wheel drive track ascends the steep northern flank of the mountains southwest of Fins to cross the Selmah plateau at a height of around 1800–2200 m to descend eventually to Wadi Tayyin. Remnant ancient stone tombs showing remarkable building architecture can be seen along the highest passes (Fig. 13.18).

Geologically the mountains south of Qalhat comprise ~ 500 m thickness of shallow marine fossiliferous limestones of the Paleocene–Lower Eocene Jafnayn Formation (Fournier et al. 2006). To the east and southeast, a deeper water facies is recorded in the Abat Formation which is up to 700 m thick. These rocks are overlain by 150 m of Rusayl Formation tidal flat sediments approximately 150 m thick and then up to 600 m thickness of Seeb Formation shallow marine limestones of Middle Eocene age (Nolan et al. 1990). The mountain massif of Jebel Bani Jabir rises to 2000 m and is bounded by two large faults, the north-south Qalhat fault to the east and the NW-SE aligned Jebel Ja'alan fault to the southwest. Both faults

Fig. 13.18 Ancient tombs on top of the Salmah plateau

were initially normal faults bounding the Abat basin during the Paleocene-Eocene, but were reactivated as reverse faults during the Oligocene-Miocene inversion and the beginning of compressional tectonics. The Qalhat fault extends from Qalhat south for over 70 km. The fault probably has a long history extending back to the Mesozoic, bounding the Permian-Mesozoic shelf carbonates and was reactivated during the Cenozoic (Fig. 13.19). Along the road from Sur to Al Kamil the fault can be seen to offset Miocene rocks along the edge of the Abat basin (Hoffmann et al. 2016). A regional NNE-SSW structural high, the Haushi–Huqf high, runs the length of eastern Oman and extends northwards into the Gulf of Oman. This structural high has been a persistent feature since at least the Permian and was onlapped by the Permian-Cretaceous shelf carbonates, The Haushi-Huqf–Jebel Ja'alan high also formed the western limit of the Batain Group thrust sheets and mélange to the east.

Fig. 13.19 Qalhat fault offsetting Miocene limestones south of Sur

13.8   Jebel Ja'alan

Jebel Ja'alan is the northern extension of the NNE-SSW aligned Haushi–Huqf
uplift, a basement uplift that runs parallel to the coast up from Dhofar to the
Sharkiyah district, east of Sur. Precambrian basement rocks are exposed in the core
of Jebel Ja'alan, overlain by shallow marine Maastrichtian–Eocene limestones and
deeper water sediments of the Abat basin. The entire Permian–Mesozoic shelf
carbonate sequence, and all the allochthonous Hawasina, Haybi and Semail
ophiolite thrust sheets are missing in this region. The eastern flank of Jebel Ja'alan
is a series of steep ESE-dipping normal faults (East Ja'alan fault) that downthrows
Maastrichtian Simsima Formation and Paleogene limestones against Precambrian
basement to the west (Filbrandt et al. 1990). The northern limit of exposure of the
Precambrian basement rocks is the North Ja'alan fault a series of WNW-ESE
striking faults that have a cumulative normal throw of 500–1000 m to the north

and a strike-slip component with horizontal sinistral displacement of up to 2000 m (Filbrandt et al. 1990).

The Jebel Ja'alan structure was a positive topographic feature during the Mesozoic and may have formed the eastern limit of the Jebel Akhdar–Saih Hatat shelf carbonates and the southwestward emplacement of the Semail ophiolite and underlying thrust sheets. The area to the east of Jebel Ja'alan is dominated by the Batain mélange a structurally complex belt that shows similar rocks as the Haybi complex of the main Oman Mountains (cherts, carbonates, ophiolite lithologies) in a mudstone matrix. A post-Middle Eocene compressional event resulted in formation of north-south trending asymmetric anticlines (Filbrandt et al. 1990). The youngest sediments in the Abat basin, a southward embayment of the Cenozoic shelf margin are Oligocene–Miocene in age, suggesting that the age of the large-scale folds is late Miocene–Pliocene (Fig. 13.20). A series of east-west aligned normal faults related to younger extension along the northern coast region may be local features along the flank of the large-scale NW-SE aligned Late Miocene–Pliocene folds.

Fig. 13.20 Large-scale folding of the Abat basin Cenozoic sediments affecting rocks of Paleocene to Miocene age

13.9   Sur–Ras al-Hadd

The city of Sur about 10 km east of Qalhat, is situated around a large, shallow lagoon open to the sea (Fig. 13.21). Sur was a major trading hub and center of the ancient dhow-building industry. Huge wooden *sambuqs* and *badan* dhows are built along the shore of the lagoon originally with valuable teak, imported from India. These ancient Omani dhows regularly traded across the Indian Ocean to Baluchistan, east to the Malabar coast of India, south to Zanzibar, Pemba and the east African coast, and even as far east as Indonesia and China. Tim Severin, a renowned British explorer led an expedition to retrace the voyage of these old trade routes from Oman to China. He had a traditional Omani dhow, the *Sohar*, built at Sur, a replica of a 9th century lateen-rigged dhow with cotton sails. With a crew of 25, the *Sohar* set sail in November 1980 and sailed over 9500 km from Sur across the Indian ocean to the Malabar coast of India, to the Lakshadweep islands, around Sri Lanka to Malaysia and across the South China Sea, arriving in Canton, China in July 1981. This voyage is described in Severin's book the *Sindbad Voyage* (1983). The *Sohar* now sits in the middle of a roundabout at Al Bustan near Muscat.

Fig. 13.21 Sur lagoon

The Sur lagoon is an important ecological habitat for birds around one of the few extensive mangrove creeks in Oman. The lagoon is home to waders and plovers including the Greater and Lesser flamingos, crab plovers, avocets, spoonbills, at least 5 species of heron, and numerous waders. During the spring and autumn migration this is a particularly important site for waders and raptors. From Sur the main road skirts around the Sur lagoon and heads east to Ras al Hadd, the northeastern point of the Arabian mainland. Another large coastal lagoon to the east (Khawr Jirama) is more barren and lacks mangroves. A smaller creek Khawr al Hajar lies immediately west of Ras al Hadd which is a barren windswept promontory, the final piece of Arabia before the Indian Ocean. The area is rich in archaeological sites, notably some recently excavated Early Bronze age settlements and fishing villages (Cleuziou and Tosi 2007). Ras al Hadd was also the site of an airstrip that was busy during the World war, acting as an important refuelling stop for aircraft bound for India and Southeast Asia.

South of Sur some impressive outcrops of spectacularly folded banded red and white cherts of the Wahrah Formation (Fig. 13.22). Radiolaria fossils have been dated as Late Jurassic to Late Cretaceous (Coniacian). These rocks are part of the Batain mélange and lie east of the Qalhat fault and the Jebel Ja'alan basement. The radiolarian cherts are associated with manganese deposits and one ancient

Fig. 13.22 Contorted folding in Permian radiolarian cherts, Batain thrust sheets, southeast of Sur

Fig. 13.23 Fe-Mn nodules in the Hawasina complex cherts of the Wahrah Formation, part of the Batain mélange southeast of Sur

manganese mine has been abandoned. Spectacular banded iron-manganese nodules (Fig. 13.23), similar to modern-day examples dredged from the deep oceans, can be found in this region. Rocks of the Batain mélange, although similar to parts of the Hawasina and Haybi complex, were derived from a separate oceanic domain to the east of Oman (Batain basin) during the Permian and Mesozoic.

13.10   Batain Coast

The Batain coastal region to the east of the Haushi-Huqf (and Jebel Ja'alan) basement high has a different and separate origin from the main Oman Mountains. The rocks along the Batain coast were originally termed the Batain mélange, consisting of many rocks of similar age and lithology as the allochthonous thrust sheets of the Hawasina and Haybi complexes in the main Oman Mountains domain west of the Haushi-Huqf arch. Detailed mapping has revealed far more structural coherence than originally thought, so the rocks are now described using the stratigraphic term Batain Group (Immenhausser et al. 1998). Rocks exposed in the Batain Group include Permian Qarari Formation limestones, Triassic-Jurassic radiolarian clays and cherts (Wahrah Formation) similar to the distal Hawasina complex, Triassic limestones containing the bivalve *Halobia*, Triassic, Middle Jurassic and Early Cretaceous alkaline pillow lavas and breccias associated with oceanic seamounts, very similar to the Haybi complex volcanics in the Oman Mountains to the west. All these rocks suggest a continuity of the Permian–Mesozoic NeoTethys ocean from the northern Oman–Gulf of Oman region around to the Indian ocean. These sedimentary and volcanic rocks were deposited in the Batain basin offshore the SE coast of Oman, and emplaced from SE to NW during the latest Cretaceous. The Batain basin is interpreted as representing a proto-Indian Ocean basin located offshore eastern Arabia and emplaced onto the Arabian margin during latest Maastrichtian–early Paleocene time (Schreurs and Immenhauser 1999).

One unique formation is the Santonian to Upper Maastrichtian (Late Cretaceous) Fayah Formation consisting of deep-water flysch-type sedimentary rocks which crops out along the entire eastern Batain coast margin but has no equivalent in the Oman Mountains (Schreurs and Immenhauser 1999). On Masirah Island the Fayah Formation sediments are sandwiched between the ophiolite thrust sheets (Immenhauser 1996). The major difference between the Batain coast to the east, and the Oman Mountains to the west of the Haushi–Huqf high, is that the Batain Group rocks or mélange was emplaced from ESE to WNW during the Maastrichtian to Paleocene, unlike the Hawasina, Haybi and Semail ophiolite thrust sheets in the Oman Mountains that were emplaced from the NNE to SWW during the Turonian to Campanian.

13.11   Asselah and al Ashkirah Lamprophyres

Around Asselah and Al Ashkirah beaches a very unusual series of highly alkaline lamprophyre, carbonatite and volcanic intrusions outcrops along the coast. Two diatremes or volcanic plugs are exposed at low tide, both at Asselah beach and Al Ashkirah beach. The rocks are composed of olivine, pyroxene and phlogopite (white mica) bearing volcanics with secondary alteration products including calcite and serpentine. These volcanoclastic rocks have accretionary lappili associated with carbonate-bearing tuffs within the Late Jurassic–Early Cretaceous Wahrah Formation. Originally thought to be kimberlites, these rocks are not true diamondiferous kimberlites, but more like hybrids between alkali basalts and lamprophyres and are associated with igneous carbonatites. Kimberlites, the source of all the World's diamonds, are intrusive pipes that are ultimately derived from the deep mantle. They are found only in regions of old, cold, stable Precambrian cratons like Africa, Canada, Australia and Siberia. The petrogenesis of the Asselah and Al Ashkirah rocks suggests a deep mantle source, possibly even a mantle plume, but later undergoing prolonged differentiation and metasomatism during ascent of the magma. Zircons from the Asselah pipe were dated at 137 Ma (Nasir et al. 2008, 2011). The Jurassic was a time of stable sedimentation during the rifting of the Tethys ocean and it is likely that these volcanic hot spots were related to similar deep mantle sources as known from oceanic islands such as the Canaries or Ascension Island in the Atlantic, or the Comores islands in the Indian Ocean. Al Ashkirah beach is also known as the best surfing beach in all Arabia, especially during the summer when big waves are associated with the *khareef* monsoon winds blowing in from the east.

13.12   Wadi Bani Khalid

Wadi Bani Khalid drains the southeastern part of the Salmah plateau and, like the Ash Shab gorge, is a region of great natural beauty with a perennial flowing river linking beautiful blue wadi pools, tree-lined river banks and some of the most spectacular waterfalls in Arabia (Fig. 13.24). The upper Wadi Bani Khalid has become a tourist destination around some beautiful pools that continue up towards a narrow cave. Some pools are deep enough to offer some great high-diving platforms. The lower Bani Khalid gorge requires a short trek and boulder-hopping to access some of the most magnificent blue pools with perennial waterfalls (Figs. 13.25 and 13.26). The wadi cuts through Paleocene–Eocene fossiliferous limestones and disgorges south into a large palm-fringed pool before channelling underground to feed underground aquifers beneath the Wahiba Sands and the Sharkiyah region. Wadi Bani Khalid is situated on the southern flank of Jebel Bani Jabir. This jebel effectively forms a single giant box fold with steep flanks and kink band sets along the northern margin and a more gently folded southern margin. A major syncline axis runs NNW-SSE along Bani Khalid with the youngest Eocene sediments exposed along the syncline axis. Folding must be post-Eocene and possibly even as young as Late Miocene–Pliocene as it also affects rocks of this age in the Abat basin to the east. The axis of this Miocene anticline is roughly ENE-WSW and at right angles to the NNE-SSW aligned folds along the Baid anticline and the alignment of the Semail Gap.

Fig. 13.24 Pools in the upper Wadi Bani Khalid

13.13 Baid Exotic

The Baid (Ba'id) Exotic is an isolated exotic massif cropping out south of Saih Hatat and in an uplifted north-south aligned late anticline axis that folds all units including the overlying Semail ophiolite. In interior Oman the Early Permian glacial Al Khlata Formation sediments are overlain by a widespread shallow marine transgression represented by the Haushi limestones of the lower part of the Gharif Formation. Lower sandstones and upper limestones represent the start of widespread rifting of the Arabian plate and was a forerunner of the major Tethyan rifting which occurred between the Early and Middle Permian. The Baid exotic is a fragment of this Late Permain margin that rifted away from the main Arabian continental margin and was subsequently caught up in the allochthonous thrust sheets of the Late Cretaceous emplacement of the Semail Ophiolite and underlying Haybi and Hawasia thrust sheets.

Fig. 13.25 Wadi pools in lower Wadi Bani Khalid

Fig. 13.26 Waterfalls in the lower Wadi Bani Khalid

The Baid Exotic consists of Upper Permian (Baid Formation) and Upper Triassic (Alwa Formation) dolomites and limestones overlain by thin red shales
indicating a drowning of the carbonate platform. The lower part of the Alwa
Formation contains the characteristic red cephalopod limestones of the Triassic
Hallstatt facies and the Cretaceous '*Ammonitico Rosso*' facies (Blendiger et al.
1990; Blendiger 1995; Pillevuit et al. 1997). The highest stratigraphic levels are
composed of Upper Jurassic–Lower Cretaceous thin-bedded limestones with

interbedded red cherts (Ta'yin Formation). The Baid Exotic is in a lower structural position in the allochthon than other typical "Oman Exotics" suggesting that it was palaeogeographically closer to the Oman continental margin during the initial Late Permian and Triassic rifting event. The importance of the Baid Exotic lies in its unique structural position within the thrust stack and the presence of the Hallstatt facies red cephalopod-bearing limestones.

References

Blendiger W, Van Vliet A, Hughes-Clarke M (1990) Updoming, rifting and continental margin development during the Late Paleozoic in northern Oman. In: Robertson AFH, Searle MP, Ries AC (eds) The geology and tectonics of the Oman region. Geological Society, London, Special Publication no. 49, pp 27–38

Dale A, Hadwin J (2001) Adventure trekking in Oman, 253 p

Filbrandt JB, Nolan SC, Ries AC (1990) Late cretaceous and early tertiary evolution of Jebel Ja'alan and adjacent areas, NE Oman. In: Robertson AFH, Searle MP, Ries AC (eds) The geology and tectonics of the Oman region. Geological Society, London, Special Publication no. 49, pp 697–714

Fournier M, Lepvrier C, Razin P, Jolivet L (2006) Cretaceous to Paleogene post-obduction extension and subsequent Neogene compression in the Oman Mountains. GeoArabia 11:17–40

Hoffman G, Meschede M, Zacke A, Al Kindi M (2016) Field guide to the geology of Northeastern Oman. Borntraeger Science Publishers

Immenhauser A (1996) Cretaceous sedimentary rocks on the Masirah Ophiolite (Sultanate of Oman): evidence for an unusual bathymetric history. J Geol Soc London 153:539–551

Immenhausser A, Schreurs G, Peters T, Matter A, Hauser M, Dumitrica P (1998) Stratigraphy, sedimentology and depositional environments of the Permian to uppermost Cretaceous Batain Group, eastern Oman. Eclogae Geol Helv 91:217–235

Nasir S, Al-Khirbash S, Rollinson H, Al-Harthy A, Al-Sayigh A, Al-Lazki A, Theye T, Massone H-J, Belousova E (2011) Petrogenesis of early Cretaceous carbonatite and ultramfic lamprophyres in a diatreme in the Batain Nappes, eastern Oman continental margin. Contrib Miner Petrol 161:47–74

Nolan SC, Skelton PW, Clissold BP, Smewing JD (1990) Maastrichtian and early Tertiary stratigraphy and palaeogeography of the central and northern Oman mountains. Geological Society, London, Special Publication no 49, pp 495–519

Pillevuit A, Marcoux J, Stampfli G, Baud A (1997) The Oman Exotics: a key to the understanding of the Neotethyan geodynamic evolution. Geodin Acta 10(5):209–238

Schreurs G, Immenhauser A (1999) West-northwest directed obduction of the Batain Group on the eastern Oman continental margin at the Cretaceous-Tertiary boundary. Tectonics 18:148–160

Severin T (1983) The Sinbad Voyage

Central Oman, Wahiba Sands and Al Wusta

Central Oman includes the vast desert areas south of the main Oman Mountain range and north of Dhofar province (Figs. 14.1 and 14.2). This area includes the major oil field provinces from west to east, the Lekhwair High, the Fahud salt basin (Natih, Fahud, Yibal, Al Huwaysar oil fields), the Ghaba salt basin, including six surface-piercing salt domes, and the Haushi-Huqf basement high. Both the Fahud and Ghaba salt basins extend northwards up to, and even beneath, the mountain front, with minor salt intrusions penetrating the Hawasina and Haybi complex thrust sheets beneath the ophiolite recently found in the Hawasina Window and at Jebel Qumayrah (Chap. 9). The Fahud and Ghaba salt basins are separated by the NE-SW aligned Makarem High. The salt is the Proterozoic–Early Cambrian Ara Group salt that extends all across Arabia and northwards to the Zagros mountains in Iran where it is termed the Hormuz salt.

Whereas the Oman Mountains show a NE-SW oriented maximum compressive stress field during the Late Cretaceous ophiolite obduction period and the Miocene, Central Oman shows an orthogonal stress field with a WNW-ESE oriented maximum compressive stress (Filbrandt et al. 2006). The SW-directed emplacement of all thrust sheets in the Oman Mountains (Hawasina, Haybi and Semail ophiolite) appears to have had little or no influence on the foreland. It has been suggested that the stress field in Central Oman was related to the WNW-directed emplacement of the Masirah ophiolite and Batain Group, or even an oblique collision of the Indian plate as it was moving northwards. The Masirah ophiolite is an older (Jurassic) segment of Indian Ocean crust, with a Cretaceous ocean island sequence built on top (Peters et al. 1997). The Masirah ophiolite is bounded by sinistral strike-slip faults running offshore the SE coast of Oman. The headlands of Ras Jibsch and Ras Madrakah both have small ophiolitic mantle sequence rocks faulted against the

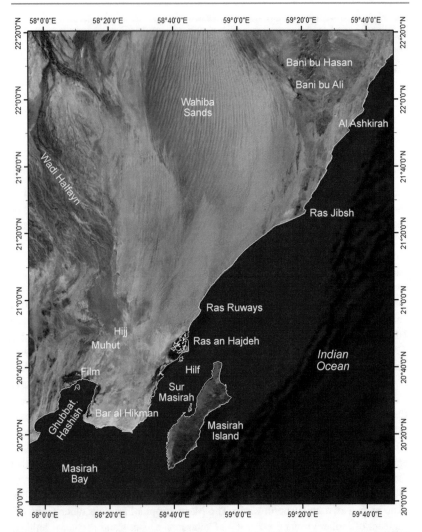

Fig. 14.1 Landsat photograph of Central Oman from the Wahiba Sands south to Masirah Bay, showing main geological locations

continental crust of Arabia. Along the south Arabian coast the transition from continental crust to oceanic crust appears very abrupt, with large-scale NNE striking sinistral strike-slip faults immediately offshore.

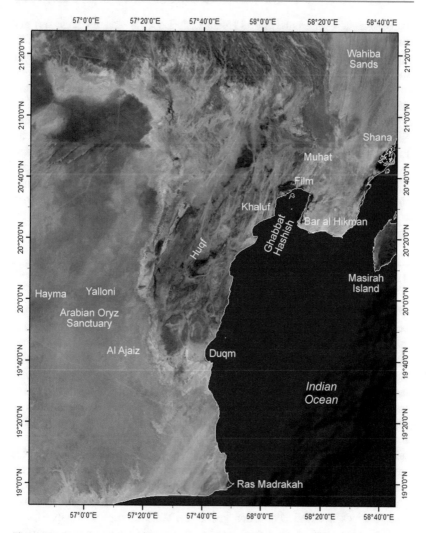

Fig. 14.2 Landsat photograph of Central Oman from Masirah Island south to Ras Madrakah showing main geological locations

Central Oman also includes the spectacular Wahiba (or Sharkiyah) sands, an isolated sandsea of north-south aligned giant seif dunes, formed by prevailing northerly winds from the Indian Ocean. South of the Wahiba sands the coast has impressive sabkhas around Bar al-Hikman. This coastline enjoys huge surf and strong winds (*kharif*) during the summer monsoon, and several shipwrecks lie along the shore, testament to the force of the monsoon. Moisture and dew from the Indian Ocean winds provides water that sustains wildlife in the desert interior, notably the Arabian oryz, gazelle, ibex, fox and ubiquitous Arabian monitor lizards (*thub*) in the inland desert.

Many unique geological sites are present in Central Oman ranging from World-class fossil sites (Jebel Saiwan rudist reefs), to sites showing evidence for Precambrian glacial events (Snowball Earth), Precambrian salt domes intruding more up to 9 km thickness of overlying rocks to pierce the surface, present-day sabkhas (Bar al Hikman), and some of the oldest known oil reservoirs in the World (Marmul, Ghaba, Huqf). Probably some of the most unique sites are those where both Lunar and Martian meteorites have been found in the desert area west of the Huqf.

14.1 Jebel Fahud, Jebel Natih

The great explorer Wilfred Thesiger was the first European known to have seen both Jebel Fahud and Jebel Natih during his trek from the Liwa oasis to Haushi in 1947–1946. He wrote '*The limestone of which they are formed had been weathered to leave no prominent features, and no vegetation was apparent on the naked rock. Both of them were dome-shaped, and I thought regretfully that their formation was of the sort which geologists associate with oil. But, even so, I did not anticipate that eight years later an oil company would have established a camp, an airfield, and be drilling at Fahud.*' In 1948 F. E. Wellings, chief geologist of IPC, scouted out the Jebel Fahud anticline from the air and decided that this was an ideal oil-bearing 'whaleback' anticline. The main topography of the Fahud anticline is the resistant boulder beds of the Eocene Umm er Radhuma Formation rubbly limestone. These shallow water limestones overlie shales of the Fiqa Formation, part of the Aruma foreland basin, which in turn overlie limestones of the main reservoir rock, the Natih Formation, part of the Wasia Group (see Chap. 3).

The western end of the Fahud anticline is a classic shallow-dipping anticline fold, but at the eastern end erosion has breached down to Late Cretaceous Fiqa shales in the core of the anticline. The first exploration oil well, Fahud-1, was spudded on 18th January 1956 in the core of the anticline, but after 15 months of drilling it was abandoned (Fig. 14.3). The well missed the reservoir by only 200 m. The oil-bearing reservoir of the Natih Formation (Wasia Group) and the overlying shales of the Fiqa Formation were later found to have been dropped down to SW

Fig. 14.3 Geologist Peter Walmsley looking over the Fahud-1 drill rig at the original site

along a large-scale SW-dipping normal fault that cuts the anticline axis (see Chap. 4). Following the first dry hole, the on-site geologists (Sheridan, Morton, Walmsley, Collumb, Melville) suggested that the rig be moved to the other side of the fault, a suggestion that, if taken up, would undoubtedly have hit oil for the first time in Oman. Most of the IPC partners withdrew following the first dry well, leaving only Shell and the Gulbenkian Foundation, the founders of Petroleum Development, Oman (PDO). In January 1964 Fahud-2 was spudded on the NE side of the fault and struck oil at relatively shallow depths of a few hundred meters. Oil is thought to have migrated into the Fahud structure both from source rocks in the Late Cretaceous Natih to the NW and from the Late Precambrian Huqf to the east about 50–40 million years ago (Terken 1999). Fahud-2 was abandoned in 2005 having pumped more than five million barrels of oil from Oman's largest reservoir (see Chap. 5).

The Fahud fault, active during the Late Cretaceous is the main structure bounding the Fahud field and has a total vertical displacement of 1 km with a minor strike-slip component. The fault does not penetrate up into the Paleocene rocks showing that the main fold growth was Campanian-Maastrichtian and part of the youngest deformation associated with the SW-directed thrust sheets in the Oman Mountains. However, the fault and fracture pattern of the Fahud, Natih, Yibal and Al Huwaisah regions does not conform to one expected from a NE-SW maximum compressive stress field as seen in the Oman mountains. A series of NW-SE or WNW-ESE aligned strike-slip faults cut the foreland region south of the mountain front. These include the Maradi fault, Natih fault and Fahud fault. Salt halokinesis was active along the faults and fractures associated with a transpressional stress field. The Maradi fault runs from the frontal ranges of the Oman Mountains SE to Masirah Bay, and shows evidence for multiple phases of motion. The WNW-ESE aligned Fahud fault moved in a dextral sense during the Late Cretaceous and then was reactivated in a sinistral sense during the Cenozoic (Al-Kindi and Richard 2014). It is still unclear what caused this NW-SE maximum compressive stress field in the foreland region extending north into Abu Dhabi.

14.2 Salakh Arch (Jebels Madmar, Salakh, Nihayda)

The Salakh arch, also known as the Adam foothills, comprises six doubly-plunging anticlines along the foreland fold-thrust belt of the Oman Mountains, all showing asymmetric anticlines above a blind (buried) SSW-vergent thrust fault. The thrust is the most southerly of all the Oman mountains thrust faults, active in the Campanian, and is the southern limit of the Late Cretaceous deformation. Jebels Salakh, Qusaybah and Nihada west of Adam village, and Jebel Madmar, east of Adam, were all formed by NNE-SSW compression during the latest part of the Late Cretaceous ophiolite obduction event. Wilfred Thesiger first noted the two dome-shaped jebels of Salakh and Madmar in the late 1940s. Two wildcat wells were drilled, Qusaibah-1 in 1969 and Madmar-1 in 1988, but neither found oil or gas in economic quantity.

North of these foreland jebels thin-skinned thrust sheets of Hawasina complex deep-water sediments overlie thin Aruma shales and Fiqa Formation with underlying shelf carbonates of the Natih Formation. The foreland fold-thrust belt includes all the overlying Hawasina complex (Hamrat Duru Group) oceanic rocks which have been carried in a piggy-back fashion above the Salakh-Madmar thrust. Seismic data shows that the thrust cuts up from pre-Permian basement through the whole Permian-Mesozoic shelf carbonate sequence, and the overlying Aruma

Group (Fiqa Formation) foreland basin sediments. At Jebel Madmar two orthogonal sets of normal faults are oriented NE-SW and NW-SE and the associated fractures provide hydrocarbon fluid migration pathways.

Jebel Madmar is an asymmetric doubly plunging anticlinal box-fold with a gently north dipping (30°) northern limb and a steep southern limb, even overturned in places. Rocks of the Cretaceous Natih and Nahr Umr Formations have been folded along with the entire Mesozoic–Permian shelf carbonate sequence beneath. The style of folding is one of fault propagation folding, and asymmetric box folding with short distance thrusts developed along one kink-band set. The thrust cuts through the Cretaceous shelf carbonates, and the overlying Late Cretaceous Muti-Fiqa Formations (Aruma Group) and the tip-line ends in the Early Cenozoic implying that the thrust along the southern margin must be Paleocene in age. Small-scale thrusts are well exposed along the northern flank of Jebel Madmar (Fig. 14.4). Two orthogonal normal fault sets are oriented NE-SW and NW-SE. Numerous joints and fractures have formed during folding and uplift of the Jebel Madmar fold.

A single isolated jebel composed of folded shelf carbonate similar to these jebels occurs at Jebel Madar to the east. Gravity surveys show that this jebel is

Fig. 14.4 Frontal thrust ramp and flat at Jebel Madar

underlain by a Precambrian salt dome at depth, the northeastern extremity of the Ghaba salt basin. Seismic sections across Jebel Madar clearly show a vertical central core of salt cutting and offsetting the surrounding rocks during upward flow. Jebel Madar, approximately 8 km by 5 km, is located approximately 40 km east of Jebel Madmar, and is also comprised of shelf carbonates ranging from upper Triassic Mahil Formation through Jurassic Sahtan Group to Lower Cretaceous Kahmah and Wasia Groups. A hiatus and major unconformity above the Albian-Cenomanian Nahr Umr and Natih Formations (Wasia Group) is overlain by shales of the Fiqa Formation, the foreland basin deposits. Fault and fracture orientations at Jebel Madar are consistent with the structures formed above salt domes (Claringbould et al. 2013).

Carbonates of the Natih and Shuaiba Formations form two major oil reservoirs in northern Oman. Source rocks, reservoir and seals can all be examined on this jebel. Seven main sedimentary cycles (A–G) are known in the Natih Formation, each consisting of a deepening and shallowing-upward sequence. Each cycle starts with a basal argillaceous layer overlain by bituminous chalks, calcareous shales and marls. Thin mixed carbonate—clastic units are overlain by thicker bioclastic carbonates, several tens of meters thick. Rudist and stromatoporoid mounds form the top of the cycle. Coarser grained carbonates are reservoirs and overlying shales form effective seals.

14.3 Umm al-Sammim

The Umm al-Sammim (Arabic for Mother of Poisons) is an inland sabkha or playa covering an area of 2400 km^2 mostly within Oman but straddling the border with Saudi Arabia (Glennie 2005). Although several wadis draining south from the Oman Mountains provide some surface water, the largest of which, Wadi Umayri occasionally floods as far south as this, the salts of the Umm al-Sammim are derived mainly from the upward flow of artesian water from the Paleocene Umm er Rahduma limestone aquifer which is here at its lowest point. The Umm al-Sammim is a shallow basin only 59 m above sea-level, bounded by small cliffs of Miocene limestone to the NW and SE. The present-day halite salt crust forms classic polygons from 50 cm thick up to 3.5 m thick, and overlays the Miocene Fars Formation bedrock of this area (Fig. 14.5). It is generally assumed that sabkha formation in Arabia started less than 7000 years ago as a result of post-glacial eustatic sea-level rise. However, recent geological and chemical data on the brines suggest that the Umm al-Sammim sabkhas date from a much older, wetter period during the Pliocene—Early Pleistocene (Fookes and Lee 2009).

Fig. 14.5 Polygons in halite salt crust of the Umm al Sammim

14.4 Wahiba Sands (Al Sharquiya Sands)

The Wahiba Sands form an isolated sandsea of long linear seif (Arabic for sword) dunes hundreds of kilometers long running from the southern margin of the northern mountains south as far as the Indian Ocean coast (Figs. 14.6 and 14.7). Individual dunes reach more than 100 m high (Fig. 14.8). The sands are generally locally derived with carbonate and ophiolite grains in the north with a slight reddish tinge indicating iron oxides, and whitish more quartz and carbonate-rich sands in the south. Bertram Thomas travelled through the western margin of the sands in 1928 and Wilfred Thesiger crossed the sands during his extensive travels in 1947. Nearly 3000 bedouin live in and mainly around the margins of the Wahiba sands, from tribes like the Janabah, Mawalik, Hikman and Sharquiya. Many of the most prized racing camels in Arabia come from this region of Oman (Fig. 14.9). Large camel farms can be seen in the northern sands, south of the larger towns of Bidayah, Al Minterib, and Al Wafi along Wadi Batha. In the south fishermen earn

Fig. 14.6 Dunes in the Wahiba Sands, Sharkiya district

Fig. 14.7 Camel train crossing the Wahiba Sands

Fig. 14.8 Individual sand dune over 60 m high in the northern Wahiba Sands

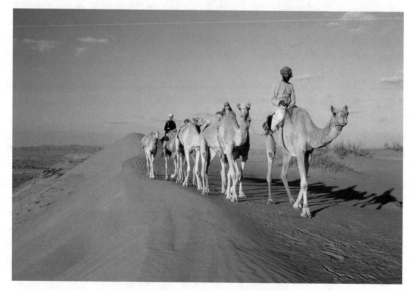

Fig. 14.9 Said bin Jabber, riding camel across the Wahiba Sands

a living from the rich fishing grounds offshore. A major geographical, geological and natural history research project studying the Wahiba Sands was conducted by the Royal Geographical Society in 1985–1987 and the results were published in the Journal of Oman Studies Special Report no 3 edited by Dutton (1988) and Winser (1989).

The linear dunes indicate strong northerly blowing winds with more variable modern wind systems superimposed on the earlier sandsea structure. The present-day climate has a weak NE winter monsoon resulting in steeper, west-facing dune slip-faces, and a strong southerly or SW summer monsoon resulting in northeastward migrating dune slip-faces. Two major wadi systems of Wadi Andam and Wadi Halfayn delineate the western boundary of the Wahiba Sands. Along the western and eastern margins of the great sandsea unique area of woodlands of acacia-type trees (*Prosopis cineraria*) called *ghaf* by the local Bedouin. Roots of the *ghaf* trees can reach more than 30 m tapping underground water reservoirs.

The Wahiba sandsea is unique in Arabia and completely separated from the Rub al-Khali sands of the Empty Quarter. There is a proposal to create a National GeoPark covering approximately 1500 km^2 including most of the big dune country in the middle and south of the region. The northern park boundary would be southwest of the major towns located along the north and eastern margins, Al Minterib, Al Kamil, Al Wafi, Bilad Bani bu Hassan and Bilad Bani bu Ali, where several tourist developments have recently sprung up. Desert camps in the middle of the Wahiba dunes have also recently opened for tourists. Some of the most geologically interesting parts of the sands lie along the south, where the modern dunes overlie the lithified fossil dunes or aeolianites.

14.5   Aeolianite Coast

Along the Indian ocean coast between Ras Jibsch and Ras Ruways, fossil dunes, aeolianites (lithified dunes with a carbonate cement) are beautifully exposed beneath the active shifting sands along the southeastern margin of the Wahiba Sands. Aeolianite cliffs showing spectacularly well-preserved dune cross-bedding, more than 15 m high occur along the coast near Ras Ruways, and may extend for more than 3 km offshore (Fig. 14.10). Aeolianite underlies most, if not all, of the Wahiba Sands making this the largest area of aeolianite in the World (Gardner 1988). The aeolianites are thought to have formed during Pleistocene high-latitude glaciations when sea-levels were lower, exposing more of the Arabian continental margin to wind erosion of carbonate-rich sand.

Fig. 14.10 Fossil sand dunes, aeolinites along southern margin of the Wahiba Sands on the Indian Ocean coast

14.6 Bar al Hikman Sabkha

Bar al Hikman is a large peninsula about 30 km long and 30 km wide along the southeast coast of Al Wusta province in central Oman. The entire peninsula is flat-lying and close to, or very slightly above, sea-level (Fig. 14.11). The interior of the peninsula which lies between 5 and 15 m above sea-level is sabkha, a mixture of sand, salt and mud, fed mainly by groundwater (Fig. 14.12). These white salt-flats are treacherous when wet. After rains the whole land is covered by a thin coating of water overlying white sabkha, a lethal cocktail for any vehicles or travellers. The coastal sabkha is regularly flooded at high tide and after heavy rains. During high tides (~ 2.8 m) the coastal areas become largely inaccessible. Low sand dunes are present along the east coast and patches of mangrove thrive near Filim in the Ghubbat Hashish bay on the western flank of the peninsula. South of the small fishing village of Al Khaluf a magnificent deserted sandy beach with white sugary dunes behind stretches for nearly 20 km along the Indian Ocean coast.

Fig. 14.11 White salt flats or sabkha, Bar al Hikman

Fig. 14.12 Thin Maastrichtian rudist-bearing limestones, Bar al-Hikman

Bar al Hikman is one of the most important sites for migratory shorebirds in Arabia and has been recognised as an internationally important wintering site for birds (Gallagher and Woodcock 1980; Eriksen and Eriksen 2005). Literally hundreds of thousands of waders that breed in the Arctic or Central Asia pass through Bar al Hikman on their migration route to Africa, while others over-winter in the coastal lagoons. Huge flocks of flamingos, together with white egrets, herons, spoonbills, avocets and the iconic crab plovers feeding on the mudflats around Mahawt Island in the Ghubbat Hashish are a magnificent sight. Along the south coast of Bar al Hikman at Khawr al Milh there are a few sandy lagoons with patches of corals in-between vast mudflats. A larger lagoon at Khawr Bar al Hikman just north of Shannah (where the ferry to Masirah Island leaves from) has extensive mangroves growing around the brackish water creeks. The low-lying sabkha regions of Bar al Hikman are an excellent modern day analogue for the shallow marine limestones on the Cretaceous Natih Formation, one of the most productive oil reservoirs in Oman. Offshore, the shallow waters are important habitats for turtles, dolphins, manta rays and whales. Seagrass beds, coral reefs and mangrove forests all offer diverse habitats for a wide variety of species.

14.7 Masirah Island Ophiolite

Masirah is the largest island off the southeast coast of Oman and a remote and fascinating place. Approximately 40 km long and 8–12 km wide it is composed of low rugged hills, ringed by extensive salt flats, sandy beaches and mudflats. A car ferry from the port of Shannur on the mainland to Hilf in the north of the island takes about 40 min. A British RAF base and airfield was closed in 1977, and a new road now encircles the island. Masirah is one of windiest places in Arabia and during the monsoon *khareef* seasons winds can reach 30–45 knots. Kite-surfing is popular in the shallow lagoons of Sur Masirah on the west coast. Masirah's beaches are also the breeding grounds for about 40% of the Worlds loggerhead turtles, with green turtles, olive ridley turtles and hawksbill turtles all nesting here between March and August. Offshore, whales can frequently be seen breaching and dolphins, rays and huge shoals of fish are common. Masirah is a very important site for migrating birds, including spectacular raptors, eagles, hawks and vultures as well as hundreds of thousands of waders.

The Masirah ophiolite exposed on the island of Masirah is a different ophiolite from the Semail ophiolite along the northern Oman Mountains. The Masirah ophiolite has both a mantle and crustal sequence, but the crust is only about 2–2.5 km thick, as opposed to 6–8 km in the Semail ophiolite (Fig. 14.13). The

Fig. 14.13 K-feldspar granites intruding the Masirah ophiolite, Masirah Island

gabbro sequence is highly condensed (only about 500 m thick) compared to the north Oman section. U–Pb zircon age dating suggests a Jurassic age of formation at 156–150 million years ago (Peters et al. 1997). The basaltic lavas are 250–500 m thick, have a MORB geochemistry and contain plagioclase, olivine, clinopyroxene and Cr-spinel phenocrysts (Peters et al. 1997; Peters and Mercolli 1998). A sequence of ocean island type alkali basalts of Cretaceous age (130–125 million years old) overlie the Masirah ophiolite (Meyer et al. 1996). Rollinson (2017) suggested that the Masirah ophiolite was a part of the Indian ocean crust, transpressionally uplifted by two strike-slip faults that occur offshore along both the west and east coasts of the island. An ancient copper mine is located in the hills about 4 km west of the abandoned village of Urf on the southeast coast of the island. Maastrichtian and Paleocene limestones unconformably overlie the Masirah ophiolite (Fig. 14.14).

Magnetic, seismic and structural data suggest that the Masirah ophiolite is part of an elongate 'basement' ridge of ophiolite that extends along the SE coastal margin of Oman (Mountain and Prell 1990). This ophiolite belt, together with two other slices of mantle rocks exposed at Ras Jibsch and Ras Madrakah, and the Batain mélange, was emplaced towards the WNW during the late Maastrichtian—

Fig. 14.14 Thin Upper Maastrichtian limestones overlying the Masirah ophiolite

early Palaeocene (Shackleton and Ries 1990; Shackleton et al. 1990; Peters et al. 1997; Schreurs and Immenhauser 1999). This timing slightly post-dates the final stages of emplacement of the North Oman Semail ophiolite belt (ca. 95–67 Ma), although the earlier stages of obduction of the Masirah ophiolite remain unknown, as the base is not exposed on land. Seismic data suggest that the Masirah ophiolite may overlie a thick mélange unit, possibly similar to the Batain mélange, and a thin Cretaceous Natih Formation shelf carbonate sequence.

14.8 Huqf Arch and Escarpment

The Huqf area contains many geological sites of international scientific importance providing fundamental geological exposures of important Palaeozoic oil and gas reservoirs (Gorin et al. 1982; Loosveld et al. 1996). These sites include two showing global glaciation deposits (NeoProterozoic and Carboniferous-Permian). Although scientific access to these sites is of crucial importance for geological research and development, many sites are in need of protection from collection and damage. The Haushi-Huqf arch is a giant NNE-SSW trending antiform that

Fig. 14.15 Giant Precambrian stromatolite mounds, Wadi Shuram, south Oman; photo courtesy of Henk Droste

exposes some of the oldest igneous and sedimentary rocks of Oman from the Neoproterozoic to the Permian. Four laccolithic type intrusions of alkaline magmatic rocks occur along the Huqf the best example exposed at Jebel Aswad. These important and unusual rocks, highly undersaturated ultramafic foidites, dated at 513–526 Ma, and some lamprophyric minettes and tephrites dated at 452–460 Ma, are probably related to a garnet and spinel peridotite mantle source and emplaced above some sort of sub-continental mantle plume (Worthing and Nasir 2008).

Classic structures such as the Khufai dome and the Buah anticline stand out clearly from the desert surface. Natural oil seeps are also present in the Gharif Formation sandstones near the Saiwan well site. The geology of the Huqf can be directly correlated to oil and gas bearing strata exposed up to 8 km deep beneath the desert to the west. Like the Oman Mountains to the north, the Huqf is uniquely important for providing geological outcrops of drilled sequences in the interior (Naylor 1996). The major geological sites of the Huqf area include:

(a) *Wadi Shuram giant stromatolite sites*

Perfectly preserved stromatolites or bacterial mats in the Wadi Shuram area provide evidence of some the earliest multi-cellular life in the Oman geological record (14.15). Oil-bearing limestone reservoir horizons containing stromatolites, sealed in massive Ara salt, are present in the sub-surface of South Oman. The exposures in Wadi Shuram are amongst the best preserved such sites in the World.

(b) *Wadi Al-Khlata glacial deposits*

The Permian-Carboniferous Al-Khlata glacial sediments in central Oman form important reservoir rocks and have therefore been studied in some detail (Levell et al. 1988; Heward 1990). Spectacular glacial pavements (Fig. 14.16) occur in the Wadi Al-Khalta and Ain Hindi regions of the Huqf where grooves scoured by ice-bound boulders are more than a meter deep (Hughes-Clark 1990). Cross-bedded sands overlie the boulders of unsorted glacial tills, rocks that are direct analogues for reservoirs that contain huge amount of heavy oil in interior Oman. The Al-Khlata Formation shows dropstones released from melting icebergs, finely laminated muds and silts (varves) and striated pavements that are amongst the best-preserved glacial deposits of this age anywhere in the World.

(c) *Permian fossil trees at Saiwan*

The Permian was a period of global warming following the Carboniferous glacial period. The Permian Gharif Formation sandstones were deposited in meandering river systems and muddy flood plains (Naylor 1996). Around the Saiwan area giant fossilised tree trunks right down to individual plant cell scale has been perfectly preserved, the petrified wood replaced by silica. Fossil wood is highly prized by collectors and it is imperative that these sites are preserved now before the specimens are all removed. One impressive large fossilised tree from Saiwan is displayed outside the Natural History Museum in Qurum.

(d) *Al Jobah granodiorite*

The only true Proterozoic basement outcrops in the Huqf are the rounded boulders of granodiorite exposed at Al Jobah. The granodiorite has been dated at 730 Ma and is thought to be the southern continuation of the Jebel Ja'alan gneisses exposed in the northeastern part of the Oman Mountains.

Fig. 14.16 Glacial pavements, Permian Al-Khlata Formation, Huqf region; photo courtesy of Bruce Levell

(e) *Qarn Mahatta Humaid*

The northern part of the Huqf in Qarn Mahatta Humaid has a 'mesa and cuesta' type of landscape. The major gas fields of central Oman occur within the Barik, Andam and Miqrat Formations, and the geological interpretations derived from outcrop observations in the Huqf have been crucial to interpretation of optimal reservoir development in the sub-surface.

14.9   Jebel Saiwan Rudist Reefs

Perhaps the most spectacular site in the whole of the Huqf is the Campanian–Maastrichtian rudist limestone reef exposed at Jebel Saiwan (Fig. 14.17). Rudists are marine bivalves related to oysters that were the main reef building organism throughout the Jurassic and Cretaceous tropical seas and died out suddenly at the Cretaceous–Paleocene (K–P) boundary, 65 million years ago. Rudists developed a variety of growth morphologies, all comprising two shells, and grew in extensive colonies that at the time were far more successful than corals. The common variety at Jabel Saiwan are tall cylinder shapes up to 60 cm high with some short thick-walled bowls up to 1 m high (*Vaccinites* sp; *Durania* sp; *Torreites* sp). Rudist limestones are very important oil reservoirs in North Oman. The Samhan Formation is the Early Campanian marine transgression across the Haushi-Huqf basement arch.

Jebel Saiwan has thousands of rudists perfectly preserved in their life positions, and is a truly remarkable site for their size, quality of exposure and their preservation (Figs. 14.18 and 14.19). The rudist reefs are laterally continuous over an area nearly 20 km wide. Despite the remoteness of the site large amounts of rudists have already been removed by collectors and it has become essential to preserve the site now and ban further collection altogether. It is now illegal to collect rudist samples from this site.

(a)

(b)

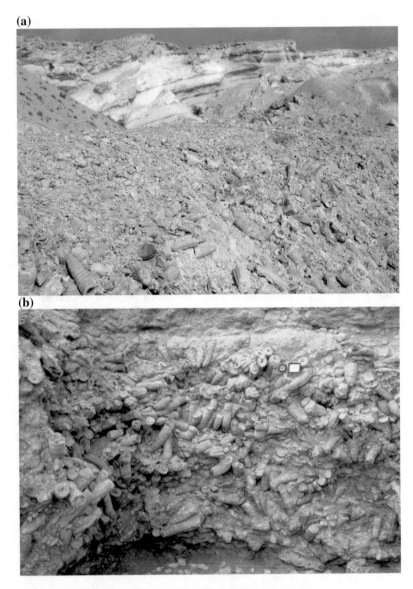

Fig. 14.17 Cliff section of Upper Maastrichtian rudists, Jebel Saiwan

Fig. 14.18 Maastrictian rudists, Jebel Saiwan

Fig. 14.19 Close-up of fossil rudists, Jebel Saiwan

14.10 Ghaba Basin Salt Domes

More than 200 Late Precambrian to early Cambrian (547–541 Ma) salt domes intrude through the complete 8–12 km thickness of Cambrian to Recent sediments across the Arabian plate. The salt (halite, gypsum and anhydrite) is termed the Hormuz salt in the Iran and Arabian Gulf region and the Ara salt in Oman. They form circular diapiric structures and are common along the Zagros Mountains, Persian Gulf region, Abu Dhabi and Oman forelands. The evaporite sequence shows classic flow banding and isoclinal folds. The rising salt plugs have carried exotic blocks or 'stringers' of many different lithologies including Neoproterozoic volcanics, Cambrian stromatolite-bearing limestones, thrombolites built of bacterial mats, sandstones, siltstones and shales. The Late Ediacaran–Early Cambrian Ara Group contains at least six evaporite—carbonate cycles capped by Cambrian sandstones of the Nimr Group and later Haima Supergroup (Reuning et al. 2009). Important oil traps occur near the salt domes.

Six emergent salt domes are scattered across the desert plains of Interior Oman in the Ghaba salt basin between the Makarem High and the Huqf region (Fig. 14.20), including Qarat al-Milh and Jebel Majayiz along the Maradi zone, and the prominent circular domes of Qarat Kibrit, Qarn Alam and Qarn Nihayda in the northern part of the Ghaba salt basin and Qarn Sahmah in the southern part (Peters et al. 2003). Gravity surveys show that several more buried salt intrusions are present at depth, some as pillow-shaped bodies detached from their roots, others as plugs intruding vertically for up to 9 km from the Ara salt horizon.

(a) *Qarat al-Milh and Jebel Majayiz salt dome*

The two northernmost salt domes, Jebel Majayiz and Qarat al Milh ('milh' is Arabic for salt) are located close to the Maradi strike-slip fault zone. Qarat al-Milh is a relatively small intrusion with only a minor topographic profile. Evaporite dissolution has resulted in erosion along a recent wadi suggesting that some of the salt intrusion is still active. Exposures of layered halite and anhydrite contain weathered blocks or stringers of carbonates with well-developed stromatolites. The larger Jebel Majayiz dome also shows evidence for recent intrusion with a north-south oriented wadi deflected around the salt dome.

(b) *Qarat Kibrit salt dome*

The Qarat Kibrit ('kibrit' is Arabic for sulphur) salt dome has a circular depression with halite, gypsum and anhydrite exposed in the core and a prominent rige along

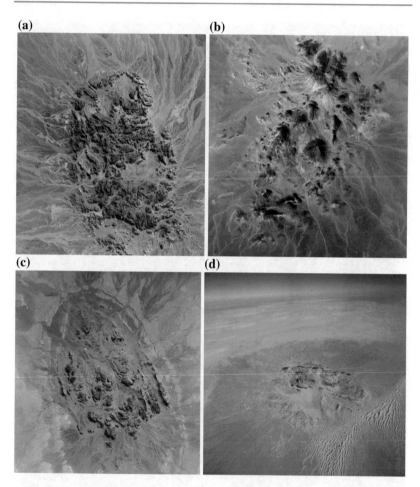

Fig. 14.20 Salt domes of central south Oman, **a** Jebel Mayjayiz, **b** Qarn Alam, **c** Qarn Nihayda, **d** Qarn Sahmah salt dome, oblique aerial photo; all photos courtesy of Henk Droste

the western margin of dark grey exotic blocks of Ara Formation bedded carbonates ('stringers') with a fetid sulphurous smell (Fig. 14.21). The sulphur is derived from the reduction of anhydrite by oil. Salt diapirism may have occurred in several phases including Late Cretaceous and post-Eocene (Peters et al. 2003). Tilted Quaternary alluvial fans around the Qarat Kibrit dome indicate that the most recent phase of salt intrusion was extremely recent. There is evidence of salt mining from the core of the dome in the past from the local Bedouin.

Fig. 14.21 Qarat Kibrit salt dome with carbonate stringers; photo courtesy Bruce Levell

Fig. 14.22 Duru tribesmen mining salt from Qarat Kibrit salt dome; photo courtesy George Laurance and Alan Heward

Fig. 14.23 Old photo (ca 1995) of Duqm beach prior to new developments

(c) *Qarn Alam, Qarn Nihayda domes*

The Qarn Nihayda salt dome forms an obvious topographic feature in the flat desert plain and is elongated along a NNW direction. Small hills composed of gypsum, anhydrite and dolomite breccias outcrop in the core of the dome with Paleogene carbonates around the rim. Stringers of Ara Formation thinly laminated fetid limestones are deformed and cut by faults. The Qarn Alam salt dome is located about 7 km SW of the Ghaba resthouse on the Muscat–Salalah highway and also shows discontinuous outcrops of Ara Formation thrombolytic carbonate stringers as exotic blocks floating in a matrix of gypsum and anhydrite.

(d) *Qarn Sahmah salt dome*

The Qarn Sahmah (Arabic for 'horn') salt dome is the farthest south salt piercement in central Oman and spanning ca 2.8 × 2.5 km is the largest, and most lithologically diverse salt diapir in Oman. It was seen and marked on maps by Wilfred Thesiger in the 1947–1948 accounts of his travels but was first studied in detail by Mike Morton, Don Sheridan and colleagues in 1956. The salt was mined by local Bani Harasis Bedouin and also by Duru tribesmen as a form of punishment; the salt was hacked out using primitive hammers and chisels then transported by camel to Ibri for barter (Fig. 14.22; Heward and Al-Rawahi 2008). The NeoProterozoic–Cambrian Ara salt has a white anhydrite cap rock breccia overlain by a thin dolomite. The salt contains several different blocks of basement granodiorite, NeoProterozoic rhyolitic volcanics, Lower Palaeozoic Haima succession lithologies and large rafts of haematised Permian Al Khlata glacial deposits, in addition to the normal carbonates with stomatolites, thromobolites and breccias (Peters et al. 2003). The Qarn Alam, Saih Nihayda and Amal fields were discovered in 1972, with Saih Rawl the following year.

14.11 Duqm Rock Garden

The stretch of coast around Duqm on the Indian Ocean coast used to be a wild and windswept coastline largely uninhabited (Fig. 14.23) The Duqm 'rock garden' is a series of bizarrely shaped cemented concretions weathered out of the Middle Eocene Dammam Formation limestones (Figs. 14.24 and 14.25). They are exposed over a wide area inland of the new port and development of Duqm. Chemical weathering, wind and rain have resulted in sculpting weird and strange shapes from these rocks. The Duqm rock gardens are now being fenced off in an effort to preserve the site as Duqm is presently undergoing massive construction around the new port and town.

Fig. 14.24 Duqm rock garden, weathered Paleocene limestone

Fig. 14.25 Paleocene limestones weathered into strange shapes, Duqm rock garden, Al Wusta region

14.12   Ras Madrakah Ophiolite

Ras Madrakah is a wild, windswept headland with beaches pounded by heavy surf during the *khareef*, and numerous shipwrecks. Offshore well-preserved coral reefs are home to a variety of marine life including shells, turtles, sharks, rays and whales. An ophiolite section outcrops on the headland at Ras Madrakah (Fig. 14.26), and together with the large Masirah Island ophiolite and more out-crops at Ras Jibsch these make up a large area of oceanic crust and upper mantle that represents older sections of Indian Ocean crust. The Ras Madrakah ophiolite consists of tectonised harzburgites and dunites of the mantle sequence and lower crustal gabbros. The upper sheeted dykes and pillow have been removed by ero-sion. Flat-lying limestones of the Oligocene (Shuwayr and Warak Formations) and early Miocene (Ghubbarah Formation) unconformably overlie the ophiolite (Fig. 14.27; Reuter et al. 2008). The Paleocene-Eocene limestones cropping out

Fig. 14.26 Mantle sequence harzburgites of the Ras Madrakah ophiolite; cliffs of Oligocene–Miocene limestones in background

Fig. 14.27 Ras Madrakah ophiolite unconformably overlain by Oligocene-Miocene white shallow marine limestones

across almost all of northern Oman and the Dhofar plateau region appear to be missing in this region. Rifting in the Gulf of Aden and southern Red Sea during the Oligocene–Early Miocene was followed by emergence of most the Arabian plate region after the Aquitanian (earliest Miocene).

References

Al-Kindy MH, Richard P (2014) The main structural styles of the hydrocarbon reservoirs in Oman. In: Rollinson HR, Searle MP, Abbasi IA, Al-Lazki A, Al-Kindy MH (eds) Tectonic evolution of the Oman Mountains. Geological Society, London, Special Publication, vol 392, pp 409–446

Clarinbould JS, Hyden B, Sarg JF, Trudgill BD (2013) Structural evolution of a salt-cored, domed, reactivated fault complex, Jebel Madar, Oman. J Struct Geol 51:118–131

Dutton RW (ed) (1988) Scientific results of the royal geographical society's Oman Wahiba Sands Project 1985–1987. J Oman Stud (Special report no. 3). Diwan of the Royal Court, Muscat, Oman, 576 p

Eriksen H, Eriksen J (2005) Common birds in Oman, Reprinted 2017. Al Roya Press and Publishing, Muscat, p 271

Filbrandt JB, Al-Dhahab S, Al-Habsy A, Harris K, Keating J, Al-Mahruqi S, Ozkaya I, Richard PD, Robertson T (2006) Kinematic interpretation and structural evolution of North Oman, Block 6, since the Late Cretaceous and implications for timing of hydrocarbon migration into Cretaceous reservoirs. GeoArabia 11(1):97–139

Gallagher M, Woodcock MW (1980) The birds of Oman. Quartet books, London, p 310

Gardner RAM (1988) Aeolianites and marine deposits of the Wahiba Sands: character and palaeo-environments, In: Dutton RW (ed) The scientific results of the royal geographical society Oman Wahiba sands project 1985–1987. Journal of Oman studies special report 3; Ministry of National heitage and Culture, Muscat,pp 75–94

Glennie KW (2005) The deserts of Southeast Arabia GeoArabia. Manama, Bahrain, 215 p

Gorin GE, Racz LG, Walter MR (1982) Late Precambrian–Cambrian sediments of Huqf Group, Sultanate of Oman. Am Asso Petrol Geol Bull 66:2609–2627

Heward AP (1990) Salt removal and sedimentation in Southern Oman. In: Roberston AHF, Searle MP, Ries AC (eds) The geology and tectonics of the Oman region. Geological Society, London, Special Publication no 49, pp 637–652

Heward AP, Al-Rawahi Z (2008) Qarn Sahmah salt dome. Petroleum Development Oman Field Guide 27

Heward A, Penny RA (2014) Al-Khalta glacial deposits in the Oman Mountains and their implications. In: Rollinson HR, Searle MP, Abbasi IA, Al-Lazki A, Al-Kindy MH (eds) Tectonic evolution of the Oman Mountains. Geological Society, London, Special Publication, vol 392, pp 279–302

Hughes-Clark M (1990) Oman's geological heritage. Petroleum Development Oman Ltd. 247 p

Loosveld RJH, Bell A, Terken JJM (1996) The tectonic evolution of interior Oman. GeoArabia 1(1):28–51

Meyer J, Mercolli I, Immenhauser A (1996) Off-ridge alkaline magmatism and seamount volcanoes in the Masirah island ophiolite, Oman. Tectonophysics 267:187–208

Mountain GS, Prell WL (1990) In: Robertson AFH, Searle MP, Ries AC (eds) The geology and tectonics of the Oman region. Geological Society, London, Special Publication no 49, pp 725–744

Naylor MA (1996) Geological features of outstanding scientific interest in the Huqf area. Petroleum Development Oman Internal Report 168

Peters T, Mercolli I (1998) Extremely thin oceanic crust in the Proto-Indian Ocean: evidence from the Masirah Ophiolite, Sultanate of Oman. J Geophys Res 193:677–689

Peters T, Immenhauser A, Mercolli I, Meyer J (1997) Geological map of Masirah North and Masirah South, scale 1:50,000. Ministry of Petroleum and Minerals, Muscat

Peters T et al (2001) Geological Map of Sur and Al Ashkharah, scale 1:100,000. Ministry of Commerce and Industry, Sultanate of Oman

Peters JM, Filbrandt JB, Grotzinger JP, Newall MJ, Shuster MW, Al-Siyabi HA (2003) Surface-piercing salt domes of interior North Oman and their significance forthe Ara carbonate "stringer" hydrocarbon play. GeoArabia 8:231–270

Reuning L et al (2009) Constraints on the diagenesis, stratigraphy and internal dynamics of the surface-piercing salt domes in the Ghaba Salt Basin (Oman): a comparison to the Ara group in the South Oman Salt Basin. GeoArabia 14:83–120

Reuter M, Piller WE, Harzhause M, Kroh A, Basi D (2008) Termination of the Arabian shelf sea: stacked cyclic sedimenatry patterns and timing (Oligocene/Miocene, Oman). Sediment Geol 212:12–24

Rollinson H (2017) Masirah, Oman's other ophiolite: a better analogue for mid-ocean ridge processes? Geosci Front 8:1253–1262

Schreurs G, Immenhauser A (1999) West-northwest directed obduction of the batain group on the eastern Oman continental margin at the Cretaceous-Tertiary boundary. Tectonics 18:148–160

Shackleton RM, Ries AC (1990) Tectonics of the Masirah fault zone and eastern Oman. In: Robertson AFH, Searle MP, Ries AC (eds) The geology and tectonics of the Oman Region. Geological Society, London, Special Publication, no 49, pp 715–724

Shackleton RM, Ries AC, Bird PR, Filbrandt JB, Lee CW, Cunningham GC (1990) The Batain Melange of NE Oman. In: Robertson AFH, Searle MP, Ries AC (eds) The geology and tectonics of the Oman Region. Geological Society, London, Special Publication, no 49, pp 673–696

Terken J (1999) The Natih petroleum system of North Oman. GeoArabia 4:157–180

Winser N (1989) The sea of sands and mists. Royal Geographical Society, London. Century Hutchinson Ltd., 199 p

Worthing MA, Nasir S (2008) Cambro-Ordovician potassic (alkaline) magmatism in Central Oman: Petrological and geochemical constraints on petrogenesis. Lithos 106:25–38

Dhofar and the Frankincense Coast

15

The southern province of Oman, Dhofar is bounded by the Indian Ocean along the south, the Yemen border in the west and the great sandsea of the Rub al-Khali or Empty Quarter to the north (Fig. 15.1). The coastal region of Dhofar is dominated by the strike-slip faults along the Arabian plate margin. The transform faulted margin was initiated in Late Jurassic–Early Cretaceous time when the African and Arabian plate separated from the India–Madagascar–Seychelles plate. The Seychelles Islands are composed of African continental crust in a small micro-plate that rifted away from East Africa and became stranded in the western Indian Ocean. Rifting may have initiated above a large mantle plume centred on the Afar region of Ethiopia and western Somaliland, which together with the Kenyan plume further south, is thought to have initiated as far back as Eocene times, around 45 million years ago. Prior to the Oligocene, the Arabian and the African plates were conjoined. Rifting of the Gulf of Aden started around 34 Ma when the continental plates spread apart, separating the Dhofar margin together with the Hallaniyat Islands (Kuria Muria Islands) from Socotra Island and the Somali part of the Horn of Africa. Oceanic spreading in the Gulf of Aden started in the Early Miocene ($\sim$17.6 Ma) and propagated northwards along the Red Sea during the Late Miocene.

The Masirah graben runs NNE–SSW along the southeast part of the Arabian plate bounded by the Haushi–Huqf basement high to the WNW and the Masirah ophiolite belt to the ESE. A major thrust fault, the Masirah Thrust, carries the Batain mélange, Masirah and Ras Madrakah ophiolite above and emplaced these thrust sheets above the Arabian shelf type sediments in the Masirah graben (Beauchamp et al. 1995). Precambrian basement rocks exposed from Jebel Ja'alan south along the Huqf arch southwards to the Hallaniyat Islands and Mirbat plains

© Springer Nature Switzerland AG 2019
M. Searle, *Geology of the Oman Mountains, Eastern Arabia*,
GeoGuide, https://doi.org/10.1007/978-3-030-18453-7_15

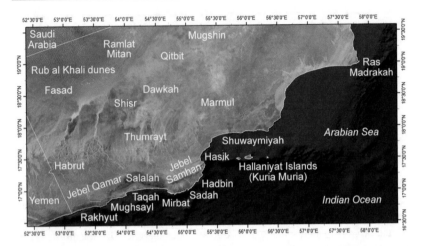

Fig. 15.1 Landsat satellite photograph of Dhaofar region, showing key locations

in Dhofar are bounded by a large east-dipping normal fault. Low hills around Mirbat expose a variety of Precambrian rocks eroded and truncated by a flat-lying unconformity. Thin Late Cretaceous sedimentary rocks overlie the Precambrian with the entire Paleozoic to Jurassic section missing. The Aptian Qishn Formation is overlain unconformably by Late Cretaceous carbonates equivalent to the Nahr Umr and Natih Formations (Beydoun 1964; Salad-Hersi et al. 2014). These Cretaceous sedimentary rocks are overlain unconformably by thick Upper Paleocene–Eocene Hadramaut Group limestones and overlying Oligocene–Early Miocene Dhofar Group limestones (Robinet et al. 2013).

Dhofar is a province of amazing contrasts. The southern coasts benefit from the full force of the Indian Ocean monsoon during the summer months with pounding surf, strong winds and rainfall. The *khareef* winds are loaded with moisture from the sea, and drop rainfall on the southern slopes of Jebel Samhan, Jebel Qara and Jebel Qamar in Dhofar, as well as the Hadramaut coast of Yemen. These hillsides are covered in verdant green grass and the vegetation has a strange Somali or African influence with baobab trees, dragons blood trees and the most valuable of all, frankincense. The fragrant incense has been tapped and collected for over 2000 years and Dhofar has the best quality frankincense (*luban*) of all. Wildlife also has an African influence notably that most magnificent animal, the Arabian leopard. The Arabian leopard used to range all over Arabia from the Northern Oman Mountains and Musandam region in the east to the Asir region of Saudi

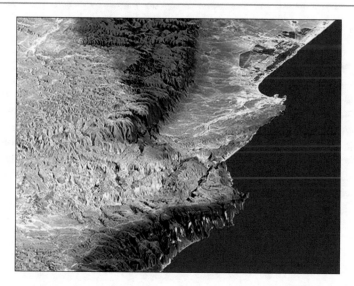

Fig. 15.2 SRTM image of the Salalah coastal plains, Jebel Qamar mountains and desert hinterland. Note the green slopes of the southeast facing cliffs, facing the brunt of the Indian Ocean monsoon

Arabia in the west, but now is confined to a few remote wadis and high mountains of Jebel Samhan with some extending west into southern Yemen. Nubian ibex, hyaena, wolf, mountain gazelle and porcupines also inhabit these mountains, whilst Arabian oryz occasional wander south from the Jiddat al Harassis. Dhofar is an important area for migrating birds, and during the autumn and spring migration, the khors and lagoons shelter hundreds of thousands of birds migrating between summer breeding grounds in Central and North Asia and wintering grounds in Africa. Rubbish dumps outside Salalah attract hundreds or thousands of storks, vultures and raptors including about seven species of Eagles.

The northern slopes of the Dhofar mountains form a great plateau that slopes gently down to north (Fig. 15.2). These are desert mountains that cannot support the woodlands and grasslands of the south despite the moisture-laden winds. The wadis here meander northwards to peter out in the vast gravel planes south of the Rub al'Khali. Wadi Andhur, north of Jebel Samhan, is a maze of barren winding canyons that are extremely difficult to navigate and easy to lost in.

15.1   Wadi Shuwaymiyah

The mountains of northeastern Dhofar are wild and rugged, with great canyons incised into the limestone plateau (Fig. 15.3). These mountains have steep flanks facing the Indian ocean and a more gentle slope westward towards the Empty Quarter. Wadi Shuwaymiyah is a spectacular eroded limestone canyon more than 30 km long, SE of Marmul and NE of Jebel Samhan (Fig. 15.4). Water seepage along bedding planes in the Paleocene–Eocene limestones drip into pools that support reed-beds providing a haven for wild life and birds. A few pools show impressive travertine dripstones with one example of a particularly impressive 'frozen' travertine waterfall. The coast here gets a real pounding from the monsoon and heavy seas make fishing almost impossible from June to September (Fig. 15.5).

Fig. 15.3 Unnamed wadi south of Wadi Shuwaymiyah cutting through Cenozoic limestones

Fig. 15.4 Pools fed by springs in Wadi Shuwaymiyah, Dhofar

Fig. 15.5 Beaches and surf on the beach at Ras Madrakah with cliffs of Oligocene–Miocene limestones in the background

15.2   Jebel Samhan

Jebel Samhan is an impressive mountain range that dominates the northwestern skyline above the Marbat coastal plains (Fig. 15.6). Jebel Samhan forms a spectacular SE or south-facing escarpment at 1500–1800 m with steep cliffs, vertical in places descending down to the Mirbat hills and coastal plain. Thin Late Cretaceous sediments overlie the Precambrian basement rocks and these are overlain by more massive cliffs of Paleocene and Eocene limestones. The central watershed of Jebel Samhan forms an incredible 'herring-bone' ridge with deeply incised wadis along both flanks and a few hidden waterholes (Figs. 15.7 and 15.8). The maze of winding wadis and steep canyon walls make trekking and navigating extremely difficult. Along the north and NW flank of Jebel Samhan the limestone dips gently away from the coast towards the interior desert. The Wadi Andhur is the largest of these wadis (Fig. 15.9). From the sea Jebel Samhan shows a gently anticline uplifting the Precambrian basement around Mirbat (Fig. 15.10). This uplift continues offshore east to the Hallaniyat Islands.

Fig. 15.6 Jebel Samhan from the south showing Proterozoic basement rocks around Mirbat overlain unconformably by thin Late Cretaceous and massive cliffs of Paleocene–Eocene limestones

Fig. 15.7 Watershed ridge of Jebel Samhan taken from helicopter. Image courtesy of Petroleum Development Oman Ltd.

These mountains and remote wadis are also home to the endangered Arabian leopard (*Panthera pardus*). Only about 20–40 individual leopards are now left in the wild. Ibex (*Capra ibex*), striped hyaena (*Hyaena hyaena*), wolf (*Canis lupus*) and mountain gazelle also inhabit these remote mountains. One of the most beautiful sights is that of the very rare and very large black Verreaux Eagle cruising along these cliffs on updrafts and swooping down to prey on rock hyrax

Fig. 15.8 Jebel Samhan escarpment

Fig. 15.9 Image of the Salalah plains and Jebel Samhan showing wadis draining north towards the Empty Quarter and the dark rocks of the Precambrian basement around Mirbat

Fig. 15.10 Jebel Samhan viewed from the Indian Ocean. Note the gentle anticline with the Late Cretaceous–Paleocene unconformity dipping gently east away from Mirbat

and other small mammals. The massive cliffs on the southern flank of Jebel Samhan are perfect for updrafts of winds coming off the Indian Ocean (Fig. 15.11).

The Jebel Samhan Nature Reserve was formed in 1997 to preserve and protect this unique wilderness area. The summer monsoon winds, the *khareef*, blows in from the Indian Ocean and these slopes are surprisingly wet and moist with dew. The conditions are just right for the growth of small frankincense (*Boswellia sacra*) trees, or *'luban'* which thrive in Jebel Samhan. The south coast of Arabia was called the Incense coast because the Worlds' best frankincense, as well as the rarer myrrh, grow here in Dhofar and Yemen. Other flora with East African origins includes the strange dragon's blood tree (Fig. 15.12), and the baobab ('up-side down') tree, which only occurs in one small wadi along the southwestern fringe of Jebel Samhan.

Fig. 15.11 Aerial photo of the escarpment of Jebel Samhan, taken from a helicopter. Image courtesy of Mateo Willis

Fig. 15.12 Dragon's blood tree, a native of East Africa growing on Jebel Samhan

15.3   Mirbat Plains Proterozoic Basement

Some of the oldest rocks in Oman are exposed along the Marbat coastal plains and low hills beneath the Paleogene limestones of Jebel Samhan. The gneisses include amphibolite facies meta-sedimentary gneisses intruded by dolerites, granodiorites and pink granite intrusions. Two major granodiorite intrusions are exposed at Marbat and Hadbin. The basement gneisses are the easternmost Pan-African gneisses in Arabia and have been cut by a swarm of rhyolitic to basaltic dykes that trend NW–SE and are particularly well-exposed around the coastal village of Sadh. The coastal stretch south of Hadbin has some beautiful sand dunes reaching over 100 m above the beach (Figs. 15.13 and 15.14). The coast between Marbat, Sadah and Hadbin has several coastal inlets that are extremely important marine sites for the mixture of tropical coral reefs and monsoon-related kelp forests. These sea-weeds can grow up to 20 or 30 m in length and provide a unique habitat in Arabian waters for a whole range of fish and shells.

Fig. 15.13 White sand beach to SW of Hadbin, Mirbat plains

Fig. 15.14 Ripples in sand dunes of white sand beach behind Mirbat

15.4 Wadi Rubkut

Wadi Rubkut is a spectacular wadi incised into the eastern flanks of Jebel Samhan
with the Proterozoic basement along the bottom and the Paleocene limestone cliffs
above (Fig. 15.15). The remote tributary canyons of Wadi Rubkut are home to the
rare Arabian leopards, Nubian ibex, wolf, hyaena, and other endangered faunas,
and therefore access to this region should be strictly controlled and limited.

15.5 Wadi Darbat Travertine Waterfall

Wadi Darbat is a spectacular wadi that drains south from the Dhofar plateau out
into Khor Rori, a large flooded wadi with extensive reed-beds forming a natural
haven for birds. The upper part of the wadi on the highlands plateau is a beautiful
stretch of blue water lined with vegetation (Fig. 15.16). Above the cliffs, a beautiful blue lagoon is flanked by green grassy banks, grazed by herds of semi-wild
camels. Wadi Darbat flows south across an incredible curtain of travertine (tufa)
rock, fossilised waterfall deposits. Inland from Khor Rori, Wadi Darbat shows an
impressive travertine curtain, a petrified waterfall over 150 m high (Fig. 15.17).

Fig. 15.15 Wadi Rubkut, taken from a helicopter. Note the Proterozoic basement gneisses in the wadi and the steep cliffs of Paleocene–Eocene limestones overlying basement. Photo courtesy of Mateo Willis

Fig. 15.16 Upper part of Wadi Darbat on the plateau

Fig. 15.17 Wadi Darbat travertine curtain

Travertine is a limestone flowstone characteristic of karst areas formed by solution and re-precipitation during extensive flooding. During the *khareef* Wadi Darbat in full spate is one of the most incredible sights in Arabia, Oman's answer to the Victoria falls. It is possible to trek to the base of the travertine curtain and also approach from top. The lower part of Wadi Darbat flows into Khor Rori, a tidal lagoon full of flamingos, herons and other bird life. Sumhuran, once the palace home of the Queen of Sheba, was a large port, flourishing on the frankincense trade before the entrance was blocked by a shingle bar. An archaeological site with old ruined buildings and a museum overlooks the lagoon.

15.6 Tawi Atair Sinkhole

Tawi Atair, the 'Well of the Birds' is one of the largest limestone sink-holes in Arabia, over 200 m deep and 150 m in diameter (Fig. 15.18). Bonellis Eagles nest in the cave and the bush around the sink-hole is alive with birds during the migration season. Another similar giant hole in the ground is the Tayk sink-hole a few kilometres north of Tawi Atair. Here it is possible to scramble down a hidden and polished path to the base of the 150 m deep hole. A new road winds down from the Jebel Samhan plateau east of Tawi Atair and a small side wadi, Wadi Hanna has about 7 or 8 baobab trees, the extraordinary shaped trees common in sub-Saharan Africa. Baobab trees have a large, fat, smooth trunk that can reach nearly 10 m in diameter with small spindly branches that look like roots, hence the nick-name 'upside-down trees'. Together with the even more extraordinary 'dragons blood trees' common along the Jebel Samhan plateau highlands, they show the influence of African species along the coast of south Arabia.

15.7 Jebel Qamar, Mughsayl, Rakhyut

West of Salalah the main coast road climbs out of the Salalah plain to the mountains of Jebel Qamar which stretch eastwards into the Hadramaut district of southern Yemen. Beautiful beaches line the coast as far as Mughsayl, where the mountains descend in spectacular cliffs direct to the sea (Fig. 15.19). Mughsayl is famed for its blowholes in the limestone coastal platform, which, during the *khareef* monsoon season can send spouts of seawater shooting 100 m up into the air, accompanied by a roaring sound as air is forced through the narrow blow-hole. From Mughsayl the road is forced inland and climbs to the 1500 m high plateau of Jebel Qamar along a series of switch-backs. The coast east of Mughsayl is

Fig. 15.18 Tawi Atair sink-hole northwest of Jebel Samhan

somewhat inaccessible with enormous sea-cliffs and small shingle beaches pounded by large surf. One small and very remote fishing village, Rakhyut is surrounded by steep cliffs and mountains but a road was recently build zig-zagging
down from the plateau (Fig. 15.20). West of Rakhyut a final small coastal village,
Dalkut, nestles beneath the towering cliffs of Jebel Qamar. Like Jebel Samhan,
Jebel Qamar has steep southern cliffs covered in green trees and wild vegetation,
fed by monsoon-lashed winds from the Indian Ocean, whilst the northern slopes
are barren deserts with a maze of meandering wadis draining north towards the
Empty Quarter. The border with Yemen cuts north in a straight line as far as the
sands of Fasad, dividing Dhofar province in Oman from the Hadramaut region of
Yemen.

Fig. 15.19 Sea cliffs at Mughsayl, west of Salalah

Fig. 15.20 The Indian Ocean coast of Dhofar, cliffs near Rakhyut close to the Yemen border

15.8 Al Hallaniyat (Kuria Muria) Islands

The rugged and remote Al Hallaniyat Islands, previously called the Kuria Muria Islands, lie 50 km offshore Hasik on the Dhofar coast. They comprise 5 islands As Sawda, Bird island, Al Hasikiyah, Al Qibliyah and Al Hallaniyat Islands. The latter is inhabited and can only be reached by sea to an anchorage at Ghubbat ar Rahib bay on the northeastern coast. They are comprised of Precambrian basement gneisses and granites similar to the Mirbat plains east of Salalah. The highest peak reaches 501 m at Ras al Hallaniyah, and all the islands have steep cliffs around them. The islands are most important habitats for oceanic birds and marine life. Whales and manta rays are commonly seen offshore with a resident population of humpback whales. For 4 months of the year during the summer monsoon the seas are incredibly rough and the islands become cut off from the mainland.

References

Beauchamp WH, Ries AC, Coward MP, Miles JS (1995) Masirah graben, Oman: a hidden Cretaceous rift basin. Am Ass Petrol Geol Bullet 79:864–879

Beydoun ZR (1964) The stratigraphy and structure of the eastern Aden protectorate. Overseas Geol Miner Resour 197

Robinet J, Razin P, Serra-Kiel J, Gallardo-Garcia A, Leroy S, Roger J, Grelaud C (2013) The Paleogene pre-rift to syn-rift succession in the Dhofar margin (northeast Gulf of Aden): stratigraphy and depositional environments. Tectonophysics 607:1–16

Salad-Hersi O, Abbasi IA, Al-Harthy A, Cherchi A, Schroeder R (2014) Stratigraphic evolution and depositional system of Lower Cretaceous Qishn Formation, Dhofar, Oman. In: Rollinson HR, Searle MP, Abbasi IA, Al-Lazki A, Al-Kindy MH (eds) Tectonic evolution of the Oman mountains. Geological Society, London, Special Publication, vol 392, pp 303–324

Rub al-Khali (Empty Quarter)

16

The Rub al-Khali desert, the largest contiguous sand desert in the World, occupies over 700,000 km^2, most of the central and south Arabian Peninsula. The central region is the great sandsea surrounded by the great Nafud desert of northern Arabia stretching as far as Mesopotamia, the Tigris and Euphrates rivers. It is bounded by rocky or stony deserts in the west in Saudi Arabia, and south in the Hadhramaut (Yemen) and Dhofar (Oman). The Rub al-Khali sand dunes stretch from Ras al Khaimah (northern UAE) in the north southwards across central and south Arabia to the mountains of Yemen and Dhofar, a distance of 1500 km (Fig. 16.1).

The main Rub al-Khali sands rise to an elevation of 1200 m and individual dunes reach heights of 250 m, interspersed with areas of sabkha, white gypsum and salt flats. The Rub al-Khali is a hyper-arid land with less than 35 mm of annual precipitation, long periods of complete drought, and temperatures reaching nearly 50 ºC (125 ºF) in summer, and as low as freezing rarely in winter.

Uplift of the western margin of Arabia began about 30–25 million years ago with the initial rifting of Arabia and Africa and the formation of the Gulf of Aden and the Red Sea. This uplift resulted in eastward tilting of the whole Arabian plate. Basement rocks are exposed along the Red Sea coasts of Yemen and Saudi Arabia, and mountains of the Hejaz and Asir regions reach over 3000 m. Proterozoic and Mesozoic rocks slope very gently eastwards all the way to the Arabian Gulf and Oman Mountains. The deserts of the Rub al-Khali mainly lie above the Mesozoic and Cenozoic sedimentary rocks of central and eastern Arabia. Along the northeastern margin of the Arabian plate foreland basins developed in front of the ophiolite thrust sheets emplaced above the Mesozoic continental margin in the Late Cretaceous and a second foreland basin developed as a result of the Arabian–Iran plate collision during the Miocene.

© Springer Nature Switzerland AG 2019

M. Searle, *Geology of the Oman Mountains, Eastern Arabia*,

GeoGuide, https://doi.org/10.1007/978-3-030-18453-7_16

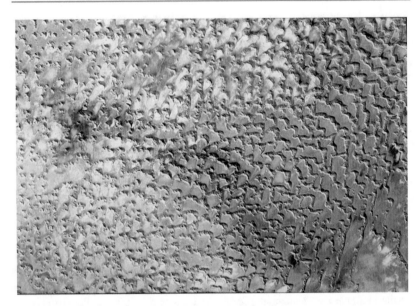

Fig. 16.1 Rolling sand dunes in the expansive Rub al-Khali desert on the southern Arabian Peninsula are pictured in this image from the Sentinel-2A satellite. The yellow-brown areas in this false-colour image are sand dunes. Shades of brown to purple reflect the mineral composition. The white patches are sabkhas where salt and gypsum lies at the surface

The Rub al-Khali dunes are dominantly affected by the southerly anticyclonic winds of the *shamal*, blowing in winter from the steppes of central Asia down the Arabian Gulf, then swinging southwest across Arabia. The western part of the Rub al-Khali has mainly linear dunes paralleling the *shamal* winds blowing south and southwest from UAE and northern Saudi Arabia to Yemen and the Hadhramaut. The eastern part of the Rub al-Khali has huge crescent-shaped barchan dunes, individually reaching a height of 250 m from Liwa oasis in the southwestern part of UAE southward in a region known as the Uruq al Mutaridah (Glennie 2005). Transverse dunes of the Empty Quarter migrate with the winds. After wind and dust storms some barchan dunes can migrate tens of meters in a single day. Plumes of sands blown off the crest of dunes avalanche down the leeward slopes. Bedouin legends talk of the 'singing sands' where cascading sand particles actually make a soft ghostly noise as they cascade down steep leeward slopes. Thesiger (1959) related "*While we were leading our camels down a steep dune face I was suddenly conscious of a low vibrant hum which grew in volume until it sounded like an aeroplane. The frightened camels plunged about, tugging at their head-ropes and looking back at the slope above us. The sound ceased when we reached the bottom*".

Between the large transverse dunes are inter-dune sabkhas indicating the present high water table over a very large area less than 100 m above sea-level. After rains the sabkhas form shallow lakes that gradually evaporate leaving white salt (halite) and gypsum to crystallise. 'Desert rose' crystals of interlocking gypsum lenses form along the interface between the sulphate-rich water table and sand dune and individual clusters of crystals can reach over 10 cm. The southern part of the Uruq al Mutaridah running along the Oman–Saudi Arabia border extends south to the Umm as Sammim. This large area of sabkha is an inland drainage basin for many of the wadis flowing southwest off the Oman Mountains, most of it sub-terranean fossil water.

In the northeast the sands lap onto the UAE-Oman Mountains, some of the sands being a distinct reddish colour (Fig. 16.2). Individual dunes show typical

Fig. 16.2 Red sand dunes of the Ramlat Sumeini lap onto the mountains of northern Oman and UAE

Fig. 16.3 A slice through a single sand dune in the United Arab Emirates showing dune cross-bedding and permanent sand-falls

dune cross bedding and sand-falls (Fig. 16.3) and can migrate with strong shemal winds. In southern Oman three great sandseas border Yemen to the west and Saudi Arabia to the north. These are the Ramlat Fasad, Ramlat Mitan, and Ramlat Barik. Gravel plains along the southeast in the Jiddat al Harasis flank the huge central area of sand dunes which make up the Rub al-Khali or 'Empty Quarter'.

16.1   Ramlat Fasad, Shisur

The Ramlat Fasad is an extensive area of high sand dunes in the northwest corner of Dhofar. These are some of the largest dunes in Oman reaching over 200 m in height and the endless barchans stretching off into the far horizon offer a magnificent sight (Figs. 16.4, 16.5 and 16.6). These dunes continue north into the heart of the Rub al-Khali. Shisur is an archaeological site (Ubar) of importance for the

Fig. 16.4 Huge dunes of the Rub al-Khali, northwest of Fasad in Dhofar

finds of Neolithic flints. Shisur was one of the most important wells for fresh water so was commonly used by the local Rashid bedouin on their raiding excursions into the Dhakaka sands in Saudi Arabia. Bertram Thomas passed through Shisur during his first crossing of the Empty Quarter in 1931, as did Philby the following year, and Wilfred Thesiger during his epic crossing of the sands in 1946–1948. An old fort at Shisur built in the sixteenth century may be the site of the 'lost city' of Ubar. During the 1970s the desert was home to herds of wild gazelle (*Dorcis dorcis*), a few rare sand cat (*Felis margarita*) and Arabian oryx (*Oryz leucoryx*). These beautiful animals have been rapidly wiped out by illegal hunting and poaching. Arabian monitor lizards, *thubs*, also used to be extremely common, and were a source of protein for the bedouin. One area in the Ramlat Fasad is notorious for its spectacular geodes measuring anything between few cm to nearly 1 m in

Fig. 16.5 Ripples on the edge of a giant sand dune, Rub al-Khali north of Fasad

diameter (Figs. 16.7 and 16.8). These circular hollow balls of rock are made of anhydrite nodules dissolved out and replaced by silica, frequently with well-formed quartz crystals growing inward to a central void. Their origin was initially as small vugs in the karst limestone, filled by silica-rich fluids which became totally enclosed and later weathered out. In the desert plains of north-western Fasad between the dunes these geodes can be found lying on the surface in their thousands.

The desert to the northwest near the Yemen border becomes more rocky with a few deeply incised wadi canyons (Figs. 16.9 and 16.10). The high dunes are lined up along the north and extend across the border into Saudi Arabia.

Fig. 16.6 Beautiful crescent-shaped dunes of the Rub al-Khali stretch north from Fasad towards the Saudi border

16.2   Ramlat Mitan

The Ramlat Mitan lies east of Shisur and the Ramlat Fasad. It is also an area of magnificent sand dunes that stretches off northwards into the Sands of Ghanim along the southern fringes of the Empty Quarter. The southern boundary of the Ramlat Mitan is a large gravel plain with buried fresh water along Wadi Atina and important water wells at Mughsin and Dawkah. Mughshin is an oasis of green in a barren desolate plain, and the site of extensive farms, reliant on irrigation. These green oases in the middle of the desert are extremely important migration stop-over points for birds. Mugshin was the last major well where Wilfred Thesiger and his band of Rashid and Bait Kathir tribesmen were able to water their camels during their crossing of the Arabian Sands north to Liwa and Abu Dhabi in 1946–7.

Fig. 16.7 A selection of silica geodes from Fasad

16.3   Ramlat Barik

The Ramlat Barik is a spectacular and beautiful region of pale orange-pink barchan dunes southeast of the Umm al Samim and north of Haima (Fig. 16.11). The dunes include large star dunes that reach heights of nearly 100 m (Glennie, 2005). It has small patches of desert plains between the dunes where acacia trees grow. Gazelle, oryz and the rare sand cats used to be common here, but now are more restricted to the Yalooni region on the Jiddat al Harassis. Golden eagles occasionally nest on isolated trees and Lanner and Saker falcons hunt for small rodents between the dunes.

Fig. 16.8 In situ geodes in Paleocene limestones in northern Dhofar. Some become weathered out to be detached and lie on the desert surface

16.4 Jiddat al Harassis

The Jiddat al Harassis is a large tract of stony desert and gravel plains that lies south and west of the Huqf escarpment and southeast of the large dunes of the Ramlat Barik and Ramlat Mitan. These desert plains are home to the Arabian oryz and the Oryx sanctuary at Yalooni was designated a National Park and UNESCO World Heritage Site, in order to preserve this iconic animal in its natural habitat. Up until the 1960s, oryx were widespread across the desert regions of south Arabia but with the advent of the 4-wheel drive, hunters and poachers, mainly from Saudi Arabia and the Gulf states, reduced the numbers to a few on the Jiddat al Harassis.

Fig. 16.9 Flat plains of the stoney desert south of the Rub al-Khali, Dhofar–Yemen border

The main culprits were private zoos that were willing to pay large sums of money to obtain a live oryz. Sultan Qaboos bin Said of Oman stepped in and funded a major international project that led to the re-introduction of wild oryz in Oman from a captive breeding herd in Arizona. The Yalooni wildlife reserve was established and for a while the oryx thrived. Rangers from the Harassis tribe kept the hunters at bay and tourists soon arrived to bolster the project. However, the number of oryz in the reserve fell from 450 in 1994 to just 65 in 2007 with just 4 breeding pairs, and as a result of the decline, protection was progressively focussed on a small fenced-in area and the size of the reserve reduced by 90%. This failure unfortunately resulted in UNESCO de-listed the site in 2007. It is also unfortunate that possible oil reservoirs are located within the larger area of Yalooni, so seismic lines have been shot across the region. There is no reason that short-term seismic

Fig. 16.10 Incised wadi on the Oman–Yemen border with a Bedouin camp

exploration (and even drilling) cannot co-exist with a wildlife reserve, and it is hoped that in the future the Yalooni oryx reserve can be expanded (even enclosing oil wells within its boundaries). It was in fact reopened as a national reserve in 2017 with oryz numbers back up to 750, so that this iconic and beautiful animal can once again roam wild across the Arabian desert.

16.5   Oman Meteorites

Meteorites are fragments of asteroids and comets that fall to Earth and are much larger versions of the meteroids (dust particles) that can be seen as 'shooting stars' as they pass through the Earth's atmosphere. Many break up and vaporise on entry, but some actually fall to Earth and are preserved as small rocks lying on the surface. Meteorites are either stony meteorites made up of silicate minerals, or iron meteorites composed mainly of iron and nickel, or combinations (stony-irons).

Fig. 16.11 Magnificent dunes with interspersed *ghaf* trees, Ramlat Barik

They give us information on the chemical composition of the solar system before the planets formed, and possibility also on the origins of Life. The Earth and Moon were subjected to millions of direct hits from meteorites during the 'Late Heavy Bombardment' approximately 3.9 million years ago. Most meteorites that fall are stony chondrites, made of mm-sized globules called chondrules in a fine-grained matrix, the oldest solid material surviving from the early solar system. Most meteorites come from the asteroid belt, but a few come from the Moon, and rare ones are derived from Mars.

The first meteorite find in Oman was in 1954 when Don Sheridan accidently ran over a meteorite whilst driving his Land Rover in Al Wusta desert region, inland from Duqm. More than 14,000 samples of meteorites have since been found in the deserts of Oman, about one fifth of all meteorites found on Earth (Fig. 16.12). In

Map of Oman with all meteorite find locations, from Al-Kathiri *et al.* **(2005)**

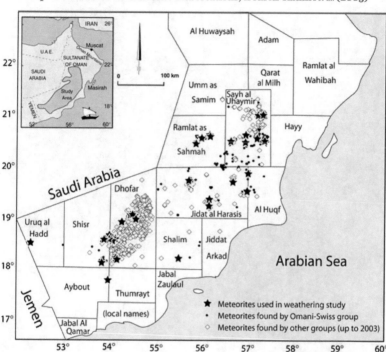

Fig. 16.12   Map of Oman with all meteorite finds, from Al-Kathiri et al. (2005)

2001 the first Martian meteorite (Sayh al Uhaymir 094) was found in Oman by a combined Omani-Swiss expedition (Fig. 16.13). Oxygen isotope compositions of minerals (olivine, pyroxene, feldspar, cristobalite) and noble gas isotopes from fluid inclusions can distinguish the origins of chondritic meteorites. Meteorites have now been found in a large swath of land from the northern Huqf region across the Jiddat al Harasis plains to northern Dhofar. It is quite likely that more meteorites remain buried beneath the sands of the Rub al-Khali. When meteorites were first discovered in Oman international collectors descended and removed many key specimens. Now the collection of meteorites in Oman is strictly prohibited by law, and all finds should be given to the Ministry of Culture and Tourism.

Fig. 16.13 Meteorites from the Jiddat al Harassis, Oman. The Sayh al Uhaymir 005 specimen is a Martian meteorite. The Ghubara specimen is a stony or chondritic meteorite

References

Al-Kathiri A, Hofmann BA, Jull AJT, Gnos E (2005) Weathering of meteorites from Oman: corrleation of chemical and mineralogical weathering proxies with 14C terrestrial ages and the influence of soil chemistry. Meteor Planet Sci 40:1215–1239

Glennie KW (2005) The deserts of Southeast Arabia GeoArabia. Manama, Bahrain, 215 p

Thesiger W (1959) Arabian sands. Longmans, 326 p

Part IV
GeoParks, Conservation and the Future

Geo-heritage and Conservation

17

Eastern Arabia includes some of the most iconic landscapes on Earth from the Oman–UAE Mountains to the vast sand deserts of the Rub al-Khali. The coastline of Musandam with its drowned fjords, steep sea cliffs and rugged mountains is unique. Several isolated sandsea regions including the Wahiba (Sharkiyah) sands, the red dunes of the Emirates, and the ancient fossilized dunes of the aeolinite coast along the Indian Ocean coastline are also unique and spectacular. The Precambrian salt domes of the Gulf islands off Abu Dhabi and those in the central Oman Fahud and Ghaba salt basins are geologically important. The most important geological sites are those along the northern mountains where the World's largest and best exposed ophiolite is magnificently preserved. These rocks show a unique profile down through 20 km of the Late Cretaceous oceanic crust and upper mantle that is by far the best preserved such ophiolite anywhere in the World. The mountains of Jebel Akhdar and Saih Hatat expose a spectacular stratigraphic profile from the pre-Permian basement rocks up through Permian and Mesozoic limestones and dolomites of the Arabian shelf carbonates. These rocks hold most of the oil and gas reserves in the Arabian Peninsula foreland, from Saudi Arabia through the UAE to Oman. The Oman-UAE mountains is the only place where these rocks are exposed, along numerous wadis that cut across the strike, where they can be studied in detail.

For all of these reasons it is imperative to preserve these geological sites for future generations. Oman has 7 UNESCO World Heritage sites including the Aflaj (*falaj*) irrigation system, archaeological site at Bat, Al Khutm and Al Ayn, Bahla Fort and the Land of Frankincense, with Qalhat added in 2018. These sites are all cultural and archaeological and all are important. This book has laid out some key geological locations that should be preserved either as World Heritage sites,

© Springer Nature Switzerland AG 2019

M. Searle, *Geology of the Oman Mountains, Eastern Arabia*,

GeoGuide, https://doi.org/10.1007/978-3-030-18453-7_17

GeoParks, or Sites of Special Scientific interest (SSSI). Several important wildlife sites have been set up as Nature Reserves in both UAE and Oman since 1976 when the Ministry of Royal Diwan Affairs passed legislation to preserve the Arabian Oryz Sanctuary at Yallooni. After that a number of protected areas and Nature Reserves were legislated by the Directorate General of Nature Conservation under the auspices of the Ministry of Environment and Climate Affairs in Muscat. These included the Daymaniat Islands, Ras al Hadd turtle reserve, and the Wadi Serin Arabian tahr reserve, all in 1996, the Al Saleel National Park, Jebel Samhan Arabian Leopard Reserve, and the Khawrs of the Dhofar coast, all in 1997.

17.1   GeoParks

It is hoped that in the near future a chain of National GeoParks will be created in order to preserve the main geological sites for the future. Figures 17.1, 17.2 and 17.3 show the rough outlines of some the more important GeoPark sites in Oman (Searle 2014). Several sites outlined in this book are most certainly of World Heritage Status, notably the Semail Ophiolite GeoPark site in Wadi Jizzi,

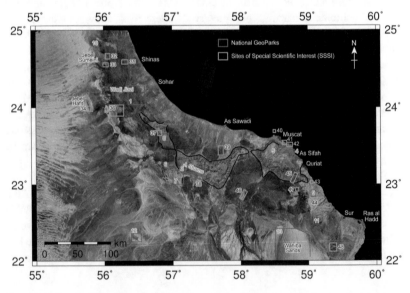

Fig. 17.1 Landsat photograph of the northern Oman Mountains showing locations of many of the proposed GeoPark sites

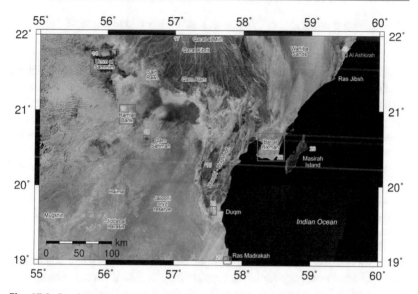

Fig. 17.2 Landsat photograph of central Oman showing locations of many of the proposed GeoPark sites

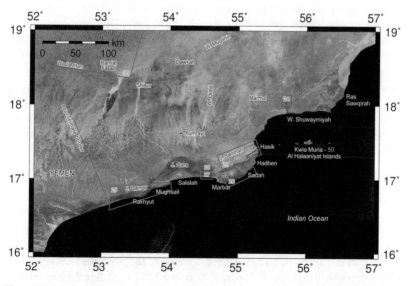

Fig. 17.3 Landsat photograph of south Oman and Dhofar province showing locations of the proposed GeoPark sites

the Musandam mountains and coastline, and the Jebel Akhdar massif. Other sites are of immense geological importance, notably the unique eclogite site at As Sifah, the World's largest and most beautifully preserved sheath fold at Wadi Mayh, the 1000 m high cliffs of the Jebel Misht and Jebel Kawr 'exotics', the Maastrichtian fossil sites at Jebel Saiwan and Jebel Sumeini, the salt domes of central Oman, the aeolinite coast and the Wahiba sands and many others. The main purpose of these GeoPark sites is to prevent destruction of the site by new developments and construction. In the past, some sites have been developed where the contractors simply did not know about the geological or geographical importance of the area. One man and a bulldozer can destroy these unique sites in a single day. It is crucial that commercial and building development companies liaise with geologists prior to any building permission being granted. New road developments are being planned now, without the involvement of any geologists. Some of these proposed routes could easily be diverted around unique geological sites.

The total land area of all the sites outlined in this book is less than 10% of the total area of UAE and Oman. None of the sites are located in major conurbation areas, or areas necessary for oil and gas drilling. Should it become necessary to drill for oil and gas within a reserve such as Yalooni or the Huqf region, there is no reason that oil exploration (seismic and drilling) cannot be combined with geo-conservation within SSSI sites. Some flexibility is required in only a few special sites. Likewise, with mining, the chromite and copper mines in the mountains were never, and are not now economic, and their development was economically pointless. One or two individuals may have made a lot of profit from mining, but the local people and the government certainly did not, and were left, in some cases, with environmental destruction. The economic and human impact on preserving these special GeoPark sites will be minimal, but the long-term benefits will be great. The advantages are that Oman and UAE will possess some of the World's most spectacular and unique geological sites that will be of untold scientific value, and will also provide a major source of local income in the post- oil and gas era.

17.2   Water Resources

Most of Arabia lies in arid or semi-arid zones that have extremely low renewable water resources. Hyper-aridity in Arabia dates from the end of the Pleistocene recharge event 26,000 years ago to the beginning of the Holocene recharge, about 9000 years ago. During the Holocene salts contained in rainfall at the surface were transported into the recharge area. Water evaporated from surface *sabkhas* leave soluble chloride and nitrate minerals on the surface and carbonate and sulphate minerals in the unsaturated zones. Isotopic data indicate that Holocene moisture is derived mainly from previously evaporated water from the Indian Ocean consistent with a summer monsoon source. ^{14}C ages of lacustrine deposits from the Rub al-Khali and travertine deposits from the Oman Mountains suggest two major wet periods, one between 30,000 and 21,000 years ago, the other from 9000 to 6000 years ago. The shift from a wet climate to a dry climate occurred, probably fairly rapidly, from about 6000 years ago. Most of the fossil groundwater is tied up in deep aquifers in the Paleogene Umm er Radhuma and Dammam Formation limestones around the periphery of the Rub al-Khali and Oman Mountains.

With the advent of the oil and gas era, the population of this region has more than quadrupled in size to approximately 9.2 million in UAE (2013 estimate, of which 1.4 are Emirati, and 7.8 million are expatriates) and 4.4 million in Oman (2016 estimate, of which 2.5 million are Omani and 1.9 million are expatriates). Both countries have seen a spiralling upward increase in population since the 1950s, a trend that is likely to continue into the future. Conventional water resources include seasonal rainfall and flooding, springs, and groundwater channelling through man-made *falaj* systems. Unconventional water resources include desalination, and treated sewage water. With the influx of oil wealth, increased standards of living and the increasing demand for water, the solution has been to build more desalination plants. The cost of desalinating water is, however, very high. New pipelines constructed in the UAE transporting oil from Abu Dhabi in the west to Fujairah in the east across the mountains to the new port at Fujairah now lie alongside pipelines bring desalinated water east to west from the Gulf of Oman across the mountains to the major cities of UAE.

The increasing demands for water in Arabia are completely unsustainable, and a major water crisis is almost certainly looming on the horizon. The main water problems in eastern Arabia now are the depletion of aquifers by excessive pumping and irrigation, saline water intrusion especially along the Batinah coast and Salalah plains of Oman, and water quality degradation. Major agricultural developments in UAE and Dhofar in particular are pumping enormous quantities of fossil groundwater from deep aquifers that are becoming rapidly depleted. The greening of the desert has major consequences for the future.

The solutions, or partial solutions, to the looming crisis in water management are difficult, but some mitigation could be along three main lines. The construction of more groundwater recharge dams in the mountains is a short-term, seasonal, local storage solution only. Changing agricultural systems toward growing more salt-tolerant plants would help. Most importantly, individual and corporate water waste needs to be reduced, and public awareness of water conservation and use should be instilled in every person, especially in schools. Much greater emphasis should be made on renewable resources, particularly solar energy, to generate electricity. It is only a matter of time until water becomes more precious than oil in Arabia.

17.3   The Past, Present and Future

Eastern Arabia has some of the most spectacular unspoilt natural wonders in the World. Fifty years ago, the great Gulf cities in the UAE, Abu Dhabi, Dubai, Sharjah, Ajman, Umm al-Qiwain and Ras al Khaimah were dusty, fishing villages with a few mud-brick houses and *barasti* houses. Now they are glimmering modern cities with skyscrapers to rival New York. The pace of development has been the fastest ever known. Fujairah and Khor Fakkan on the east coast have become major ports and development there has also been very rapid. In Oman, the tiny capital city of Muscat is located in one of the most picturesque settings with its spiky mountains of peridotite all around (Fig. 17.4). Muscat cannot and should not expand, nor the twin city port of Muttrah (Figs. 17.5, 17.6). Major developments have spread to the Ruwi valley, and along the Batinah coast to Azaiba, and Seeb. The Qurum beach area in 1970 had not a single building in view (Fig. 17.7); now it is a metropolis encompassing Qurum heights, Medinat Qaboos, and Azaiba. In 1970 when Sultan Qaboos acceded to the throne, he landed at a dusty airstrip in the Ruwi valley in which there was only one building, the Army fort at Bait al-Falaj.

The mountain regions of UAE and Oman have, however largely remained pristine (Fig. 17.8). Networks of new roads have made the mountainous regions much easier to access, and the major towns in the interior of Oman, such as Ibri, Bahla, Nizwa, Izki, Ibra, and Sanaw have also developed rapidly. Old mud brick houses in Nizwa (Fig. 17.9) have been replaced by modern constructions. The old forts at Nizwa and Bahla (Fig. 17.10) have been restored in an aesthetic manner to preserve as much of their old character as possible. Many of the remoter wadis

Fig. 17.4 Muscat seafront from Fort Mirani, taken in 1968. Note the old flag of the Sultanate, prior to 1970

Fig. 17.5 Muttrah harbour and the old fort, with several ancient wooden dhows in foreground; photo taken ca. 1976

Fig. 17.6 Ancient *sambuq*, an old wooden dhow, probably an ancient pearling dhow, beached on Muttrah beach, photo taken ca. 1973

Fig. 17.7 Qurum beach and the mangrove creek along Wadi Aday; photo taken in 1968. Note the sand dunes behind the mangroves, and the long sand dune ridge at Bausher in the distance

Fig. 17.8 Ariel photo taken over Wadi Semail looking towards Jebel Nakhl and the Semail Gap, photo taken in 1978

Fig. 17.9 Old mud-brick city of Nizwa, taken from the circular fort, photo taken ca. 1975

around Jebel al-Akhdar are too steep and mountainous for development and they should be preserved intact (Figs. 17.11, 17.12 and 17.13). Quite apart from the geological importance of the Jebel al-Akhdar area, the incredible gorges and canyons of the wadis draining Jebel Shams and the Saiq plateau, and the magnificence of a sunrise over the Oman Mountains from the summit of Jebel Shams remains a sight of awesome beauty (Fig. 17.14). Some geological sites such as the Wadi Mayh sheath fold (Fig. 17.15) are in danger of being destroyed through road construction projects. The Wadi Mayh gorge is strongly affected by floods, when the canyon can fill up in a few minutes. Each rainstorm or cyclone destroys the road which has to be rebuild every time. It makes no sense to construct a road along this section of the wadi, and there are other more suitable sites to the east. Likewise, the eclogite beach at As Sifah is also a prime site for beach-front development. The new harbour, town and hotel site at As Sifah village are build

Fig. 17.10 Bahla fort dominating the oasis of date palms, prior to its restoration, photo taken ca. 1976

and are situated in a suitable location, but the eclogite beach to the north must be preserved intact. These two geological sites are of World Heritage importance without doubt. Many GeoPark sites require no infrastructure, indeed such development would detract from the site. Some however, especially the fragile fossil sites at Jebel Sumeini and Jebel Saiwan need rangers to look after the sites and prevent over-collection.

The desert regions of Oman and UAE have also developed to some extent, mainly around the population centers such as Adam and Sinaw, and along the Muscat to Salalah highway. Some localities such as the magnificent Wahiba sands (Fig. 17.16) should not be subjected to any developments. Their preservation for scientific, natural history and tourist recreation is reason enough to leave them in the pristine state they are now in. Likewise, with some of the more beautiful desert

Fig. 17.11 The Grand Canyon of Wadi Nakhr, view looking southwest towards Jebel Kawr from the canyon rim

Fig. 17.12 Tiny village house for goat shepherds along the rim of Wadi Tanuf

Fig. 17.13 Terraces of alfalfa grass growing in a remote village in the upper canyon of Wadi Tanuf

Fig. 17.14 Ridge upon ridge of mountains stretching into the distance from the summit of Jebel Shams (3009 m), taken at dawn

Fig. 17.15 Spectacular circular folds in Permian shelf carbonates, Wadi Mayh sheath fold

Fig. 17.16 Wahiba Sands—the sea of sands and mists

regions of the interior and Dhofar, the Barik sand dunes, and the great dunes of Ramlat Fasad in the Rub al-Khali, should be preserved at all costs. Most oil company activity, in terms of seismic exploration or well drilling, can co-exist perfectly happily with geo-conservation and some account needs to be taken of this. The oil companies have on the whole been operating in UAE and Oman in a very responsible manner, again thanks largely to excellent management.

Development in Oman has been impressive, due mainly to intelligent government and directives from the highest levels. It has now reached a stage where uncontrolled development could easily take over, mainly through beaurocratic

Fig. 17.17 Sunlight bursts through the mountain ridges and canyons of Jebel Shams

impasse, and government ministries not being joined-up. Town and country planning on a national scale are required more than ever, and the involvement of geologists, naturalists, archaeologists and other specialty scientists are absolutely necessary at the initial planning stage of all new developments. It is hoped that this book has outlined at least some of the most important geological sites in both UAE and Oman, and that this will help planners in the future to avoid these areas for any new developments. The scientific importance, the importance to wildlife conservation and the beauty of all these sites (Fig. 17.17) is too important, and we should all make every effort to conserve the natural environment and all these geological sites for future generations (Fig. 17.18).

Fig. 17.18 (end-photo). Sunburst above the Jebel Shams ridge

Reference

Searle MP (2014) Preserving Oman's geological heritage: proposal for establishment of World Heritage Sites, National GeoParks and sites of special scientific interest (SSSI). In: Rollinson HR, Searle MP, Abbasi IA, Al-Lazki A, Al-Kindy MH (eds) Tectonic evolution of the Oman Mountains, vol 392. Geological Society, London, Special Publication, pp 9–44

Appendix A

Glossary of Geological Terms

Aa	Blocky lava.
Alkali Basalt	Basalt rich in alkali elements.
Allochthonous	Formed not in the present setting, includes thrust sheets.
Ammonite	Marine coiled nautilus-type animal (Cephalopod).
Amphibolite facies	Medium pressure-temperature metamorphic facies.
Andesite	Volcanic rock formed of feldspar, hornblende, pyroxene, quartz.
Ankaramite	Type of alkali basalt, typically containing nepheline.
Anticline	Compressional fold arched upward.
Authochthonous	Formed in situ.
Asthenosphere	Deep region of Earth's mantle below the lithosphere.
Basalt	Fine-grained igneous rock, formed by volcanic eruptions.
Bastite	Alteration mineral after orthopyroxene.
Batholith	Large-scale granite intrusion.
Benioff zone	Subduction zone.
Benthic	Deep ocean environment.

© Springer Nature Switzerland AG 2019

M. Searle, *Geology of the Oman Mountains, Eastern Arabia*,

GeoGuide, https://doi.org/10.1007/978-3-030-18453-7

Biostratigraphy	Stratigraphy defined using fossils.
Bivalve	Two-shelled marine animal (Lamellibranch).
Blueschist facies	High-pressure, low-temperature metamorphic facies.
Boninite	Volcanic rock, high-Mg andesite, formed in island arcs.
Brittle	Upper crust fracturing of rocks resulting in faulting and earthquakes.
Carpholite	Single chain silicate mineral characteristic of high-pressure, low-temperature metamorphism in mudstone rocks.
Cenozoic	Period of Earth history between 65 and ~ 1 million years ago.
Chert	Siliceous sedimentary rock.
Core	Innermost part of the Earth, composed of iron and nickel. Inner core is solid; outer core is liquid.
Crinoid	Sea lily, or sea-fan, related to starfish.
Crust	Upper part of the lithosphere, 3–7 km thick in oceans; 30–70 km thick in continents.
Cumulate texture	Igneous texture composed of mafic cumulus crystals in inter-cumulus liquid phase, usually plagioclase.
Denudation	Erosional removal of rock.
Destructive plate margin	Compressional plate margins with subduction zones where oceanic lithosphere descends back into the mantle.
Diapir	Upward intrusive body of granite, or salt.
Distal	Far away from a continental margin.
Dolerite	Medium-grained basaltic rock between gabbro and basalt.
Dolomite	Sedimentary rock composed mainly of magnesium carbonate.
Ductile	Plastic, flow shear typical of higher temperature deformation beneath the brittle crust.
Dunite	Ultramafic rock composed entirely of olivine.
Dyke	Long, linear, vertical intrusive structure, sheeted dykes.
Echinoid	Sea urchin.

Eclogite facies	High-pressure, low-temperature metamorphic facies.
Erosion	Removal of rock by chemical or mechanical weathering.
Exhumation	Movement of a rock towards the Earth's surface by a combination of tectonic forces uplifting rocks and removal of overburden by erosion.
Exotic	Not formed in situ.
Facing direction	Way-up along a fold axial plane cleavage.
Fault	Fracture line in Earth's crust; sudden motion results in earthquakes.
Felsic	Light-coloured igneous rocks composed mostly of quartz, feldspars.
Foliation	Penetrative, planar fabric in rocks, e.g. cleavage, schistosity.
Foraminifera	Single-celled organism, formed of calcium carbonate.
Foreland	The area in front of a mountain belt (SW of Oman-UAE mountains).
Foreland basin	A sedimentary basin formed adjacent to a load (e.g. mountain belt).
Gabbro	Coarse-grained igneous rock composed of mafic minerals and plagioclase feldspar.
Gastropod	Coiled mollusc shell.
Geode	Hollowed rock lined inside with quartz crystals.
Geochronology	Study of dating age of rocks, commonly using radioactive decay.
Geomorphology	Study of surface landscapes and processes; geography.
Gossan	Mineralised ore deposit.
Granites	Intrusive igneous rock composed of quartz, feldspars and mafic minerals.
Granulite facies	High-temperature, medium or low-pressure metamorphic facies.
Greenschist facies	Low- to medium pressure-temperature metamorphic facies.
Gneiss	Metamorphic rock showing ductile foliation.
Gypsum	Hydrated calcium sulphate mineral.
Halokinesis	Salt tectonics.

Harzburgite	Ultramafic rock composed of olivine and orthopyroxene (enstatite).
Hinterland	The area behind a mountain belt (NE of Oman-UAE mountains).
Hotspot	Zone of upwelling anomalous high-temperature asthenosphere usually beneath ocean islands like Hawaii, Réunion island, etc.
Island Arc	Line of volcanic islands located above a subduction zone and trench.
Isobar	Line of equal pressure.
Isograd	Line of constant metamorphic grade, used for mapping; e.g. garnet-in isograd.
Isostasy	Principal of buoyant crust floating above a denser fluid-like substrate (mantle) according to Archimedes Principle.
Isotherm	Line of equal temperature.
Karst	Weathering of limestone, forming caves.
Leucogranite	Igneous rock derived from melting of continental crust.
Lherzolite	Ultramafic rock composed of olivine, orthopyroxene, clinopyroxene.
Limestone	Sedimentary rock composed of calcium carbonate.
Listvenite	Silica metasomatised serpentinised peridotite.
Lithosphere	Earth's crust and upper mantle, usually about the upper 100 km.
Mantle	Layer of Earth's interior below the crust composed of ultramafic rocks.
Marble	Metamorphosed limestone.
Mélange	Mixture of rocks, blocks in a matrix of either serpentinite (tectonic mélange) or sedimentary (shale, mudstone).
Mesozoic	Period of Earth history between 250 and 65 million years ago.
Metamorphism	Change of minerals, rocks by increase of temperature and pressure.
Metasomatism	Fluid-related alteration of rocks.
Mid-Ocean Ridge	A narrow zone of basalt volcanic rocks formed along a spreading center (e.g. Mid-Atlantic Ridge, East Pacific Rise).

Migmatite	High-grade, partially molten metamorphic rock composed of leucosome (melt phase) and restite (mafic minerals).
Moho	Mohorovicic discontinuity, geophysical boundary between the crust and mantle.
Monsoon	Regional tropical wind system usually associated with heavy rainfall, derived from Arabic *mausim*.
Mudstone	Fine-grained sedimentary rock.
Nappe	Alpine term for large-scale overfold, sometimes with a thrust at base.
Normal fault	Extensional fault placing younger, shallower rocks above older, deeper rocks.
Obduction	Process of emplacing an ophiolite complex onto a continental margin.
Ophiolite	Sequence of oceanic upper mantle peridotite and crustal gabbros, sheeted dykes, pillow lavas and deep-sea sediments emplaced onto a continental margin.
Orogeny	Mountain building episode of geological time.
Pahoehoe	Ropy lava.
Paleozoic	Period of Earth history between 550 and 250 million years ago.
Passive continental margin	Aseismic continental margin (Atlantic type).
Pegmatite	Coarse-grained granite, intrusive dyke rock.
Pelagic	Open ocean environment, far away from a continental margin.
Pelite	Metamorphosed shaley rock.
Peridotite	Ultramafic rocks composed only of mafic minerals (olivine, pyroxene, amphibole, etc) without feldspar or quartz.
Pillow lava	Rounded shape of basalt when erupted under water.
Plagiogranite	Igneous rock composed mainly of plagioclase.
Plate tectonics	Theory whereby Earth's rigid lithospheric plates move independently driven by mantle convection.
Pluton	Large intrusive body of igneous rock e.g. granite.

Pressure-temperature-time (PTt) path	Evolution of a metamorphic rock through time during its heating and burial (prograde path), and cooling and exhumation (retrograde path).
Proximal	Near a continental margin.
Psammite	Metamorphosed sandstone.
Quartzite	Sedimentary rock composed almost entirely of quartz.
Radioactive decay	Spontaneous decay of an atom to an atom of a different element by emission of one or more particles or photons from its nucleus (alpha, beta, gamma decay).
Radiolaria	Small, single celled marine plankton.
Rare Earth Element (REE)	Elements with atomic numbers between 57 and 71 occurring in very small amounts in minerals.
Rheology	Flow and ductile deformation of rocks, including viscosity, plasticity, and elasticity.
Rodingite	Calcium-rich metasomatic gabbro.
Rudist	Large, solitary coral, became extinct at the K-P boundary.
Sabkha	Flat, salt-encrusted depression.
Sandstone	Sedimentary rock with high quartz content.
S-C fabrics	Shear-related mineral fabrics (schistosity–shear-cisaillement).
Sea-floor spreading	Process by which new ocean crust is formed along Mid-Ocean ridges.
Serpentinite	Rock composed of hydrated olivine or serpentine.
Shale	Fine-grained, fissile sedimentary rock formed from compaction of mudstone.
Strike-slip fault	Vertical fault where rocks have moved horizontally.
Subduction zone	A narrow zone along which lithospheric plate descends to the mantle, usually a zone of deep earthquakes.
Suture zone	A zone of collision between two tectonic plates.
Syncline	Fold where strata have been downwarped.
Thermobarometry	Determination of the pressure (P) and temperature (T) conditions of a metamorphic rock using equilibrium thermodynamics.

Thrust fault	Compressional fault where the hangingwall is thrust over the footwall.
Turbidite	Sedimentary rock typically showing graded bedding and alternation of sandstones or limestones with shales, a result of sediment transport down a continental slope.
Transform fault	Oceanic strike-slip fault along which the tectonic plates move horizontally.
Transpression	Tectonic component of compression combined with strike-slip faulting or shearing.
Transtension	Tectonic component of extension combined with strike-slip shearing or faulting.
Trench	Long, narrow, deep bathymetric depression coincident with subduction zone along a plate boundary (e.g. Mariana trench).
Ultramafic rock	Rock composed entirely of mafic minerals.
Unconformity	Horizon representing a time gap with difference of bedding angle.
Umber	Metalliferous (usually Fe, Mn) deep-sea sediments associated with concentrated mineralisation.
Vergence	Sense of overturning in a fold.
Volcanic arc	Chain of volcanic islands above a subduction zone characterised by basalt and andesite basalts.
Websterite	Ultramafic rock composed mainly of pyroxenes.
Wehrlite	Ultramafic rock composed of olivine and clinopyroxene.
Within-Plate Basalt	Basalt formed in an oceanic island location away from a continental margin or mid-ocean ridge.
Xenolith	Fragment of country rock preserved within a volcanic or granite intrusive rock.

Appendix B

Glossary of Arabic Words

Arabic	English
Abiad	White
Akhdar	Green
Aswad	Black
Ayn	Fresh water spring
Badan	Small wooden fishing dhow
Bahar	Sea
Bait	House
Baiza	Currency, 100 baiza = 1 Rial
Baluchi	Tribe from southern Pakistan, Baluchistan
Bandar	Port
Bani	Sons of …, tribal
Barasti	Palm frond hut
Bedouin (bedu)	Nomads
Bilad	Village
Bin	Son of …
Birkat	Water well
Boom	Large sailing vessel, pearling ship
Burj	Tower
Bustan	Garden
Dhow	Sailing vessel
Dirham	Currency of the United Arab Emirates

© Springer Nature Switzerland AG 2019
M. Searle, *Geology of the Oman Mountains, Eastern Arabia*,
GeoGuide, https://doi.org/10.1007/978-3-030-18453-7

Falaj	Water course for irrigation
Hamra	Red
Hayl	Plateau
Hejaz	Western part of Arabia
Imam	Muslim leader
Jazirat	Headland
Jebel (jabal)	Mountain
Jiddat	Plain
Khanjar	Omani curved dagger, worn by men
Khareef	Monsoon winds from the east in South Arabia
Khor (khawr)	Tidal inlet
Luban	Frankincense
Magreb	West
Majlis	Village meeting place
Matara	Rain
Medinat	Town
Milh	Salt
Mina	Port
Nabkha	Sand mounds in a sabkha
Nefud	Sand desert of northern Arabia
Nejd	Desert of Central Arabia
Qarat	Salt dome
Ramlat	Sands
Ras	Headland, cape
Rial	Omani currency
Rub al-Khali	Empty Quarter central Arabia
Sabkha	Gypsum, salt-pan
Sambuq	Sailing ship used by pearl divers
Shaheen	Falcon
Shamal	Wind from the north
Shams	Sun
Sharkiyah	East
Shasha	Small Omani fishing boat made from reeds
Sheik	Tribal leader
Suhaili	Wind from the south
Sultan	Ruler of Oman
Swahili	Language of East Africa and Zanzibar
Souq	Arab market place
Tawi	Well

Thaler	Maria Theresa dollar (old currency of Arabia)
Thub	Monitor lizard
Ubar	Lost city, 'Atlantis of the sands'
Umm al-Sammin	'Mother of Poisons' (quicksand)
Uruq	Chain of linear sand dunes
Wadi	Dry river valley
Waliyat	Administrative region in Oman
Wazir	Government advisor.

Bibliography for Arabian Exploration and Travels

Allfree PS (1967) Warlords of Oman. Robert Hale, London, 191 p.
Alston R, Laing S (2012) Unshook till the end of time. Gilgamesh Publishing.
Belgrave SC (1960) The pirate coast. Libraire du Liban, 200 p.
Blanford WT (1872) Records of the geological survey of India.
Boustead H (1974) The wind of morning. Chatto & Windus, London, 240 p.
Carter John (1982) Tribes in Oman. Peninsula Publisher, London
Costa PM (1991) Musandam. Immel Publishing, London, 249 p.
Cox P (1925) Some excursions in Oman. Geogr J.
Cox P (1929) Across the Green Mountains of Oman.
Curzon Hon GN (1892) Persia and the Persian Question.
Dickson HRP (1949) The Arab of the desert. George Allen & Unwin, London, 664 p.
Doughty C (1888) Travels in Arabia Deserta. Cambridge University Press.
Eccles GJ (1927) The sultanate of Muscat and Oman. J Cent Asiat Soc 14:19–42
Facey W, Grant G (1996) The Emirates by the first photographers. Stacey International, 128 p.
Glennie Kenneth (2005) The desert of Southeast Arabia. GeoArabia, Bahrain
Henderson E (1988) This strange eventful history. Motivate, 242 p.
Lawrence TE, Seven pillars of wisdom.
Lees GM (1928) The physical geography of South-Eastern Arabia. Geogr J 71:441–466
Miles S (1884) J Excursion Oman, Southeast Arabia.
Miles S (1919) The countries and tribes of the Persian Gulf.
Miles S (1920) On the border of the great desert: a journey in Oman.
Monroe E (1973) Philby of Arabia. Faber & Faber, London, 334 p.
Morton M (2006) In the heart of the desert. Green Mountain Press, 282 p.
Morton M (2013) Buraimi: the struggle for power, influence and oil in Arabia. I.B.Taurus, London, 286 p.
Morris J (1957) Sultan in Oman. Faber & Faber, London, 140 p.
Palgrave WG (1865) Narrative of a year's journey through Central and Eastern Arabia.

© Springer Nature Switzerland AG 2019
M. Searle, *Geology of the Oman Mountains, Eastern Arabia*,
GeoGuide, https://doi.org/10.1007/978-3-030-18453-7

Paxton J (unpublished) History of PDO. Middle East Archive Center, St-Anthony's College, Oxford.

Petroleum Development Oman (2009) Oman faces and places. Muscat, Oman

Peyton WD (1983) Old Oman. Stacey International, London, 128 p.

Phillips Wendell (1967) Oman—a short history. Longmans, London

Searle P (1979) Dawn over Oman. George, Allen & Unwin, 146 p.

Searle M (2013) Colliding continents. Oxford University Press, 438 p.

Skeet I (1974) Muscat and Oman, the end of an Era. Faber and Faber, London, 224 p.

Sheridan D (2000) Fahud, the Leopard Mountain. Vico Press, Dublin, 268 p.

Stanton-Hope WE (1951) Arabian adventurer.

Stark Freya (1936) The Southern Gates of Arabia. John Murray, London, 327 p.

Stiffe A (1897) Capt. Ancient trading centres of the Persian Gulf' Geogr J.

Thesiger Wilfred (1948) Across the empty quarter. Geogr J 111:1–21

Thesiger W (1959) Arabian sands. Longmans, 326 p.

Thesiger Wilfred (1964) Marsh Arabs. Longmans, Green & co, 233 p.

Thesiger Wilfred (1979) Desert. Collins, Marsh and Mountain, 304 p.

Thesiger W (1987) The life of my choice. Collins.

Thesiger W (1987) Visions of a Nomad. Harper Collins, 224 p.

Thesiger W (1998) Among the mountains, travels through Asia. Harper Collins, 250 p.

Thesiger W (1999) Crossing the sands. Motivate, 176 p.

Thesiger W (2001) A vanished world.

Thomas Bertram (1931) Alarms and excursions in Arabia. George Allen & Unwin, London, 316 p.

Thomas B (1932) Arabia felix, across the empty quarter of Arabia. Jonathan Cape, London, 397 p.

Villiers AT (1952) The Indian Ocean.

Ward P (1987) Travels in Oman. The Oleander Press, Cambridge, England, 572 p.

Wellstead JR (1837) Narrative of a journey into the interior of Oman. J R Geogr Soc Lond.

Wellstead JR (1838) Travels in Arabia, vol 1.

Wilkinson J (1087) The imamate tradition of Oman. Oxford University Press.

Wilkinson John C (1991) Arabia's frontiers: the story of Britain's boundary drawing in the desert. I.B. Taurus, London, 400 p.

Winser Nigel (1989) The sea of sands and mists. Century, London, 199 p.